THE LIVING WATERS

Part 1

Seas and Oceans

by Gillian Standring

ISBN: 84-382-0015-X. Dep. Legal: S.S. 601-1975
Printed and bound in Spain by
T.O.N.S.A., and Roner S.A.,
Crta de Irun, Km.12,450, Madrid–34

Series Coordinator Geoffrey Rogers
Series Art Director Frank Fry
Design Consultant Guenther Radtke
Editorial Consultant Donald Berwick
Series Consultant Malcolm Ross-Macdonald
Art Editor Douglas Sneddon
Editor Allyson Fawcett
Copy Editors Maureen Cartwright
Damian Grint
Research Jonathan Moore
Julia Hutt
Art Assistant Michael Turner

Contents: Part 1

Editorial Advisers

Foreword by David Attenborough

The sea is in our blood. That statement may be almost literally true. The relative proportions of certain salts in our blood are almost exactly the same as they are in the sea. They are much more dilute, but then the sea itself was similarly dilute some 400 million years ago. That was the period when the first backboned animals were developing and changing their circulatory system from one that connected directly with the surrounding water to one that was enclosed within the body. So that may have been the time when the chemical constitution of vertebrate blood was first fixed; and it has remained similar ever since.

Life originated in the sea. Gillian Standring, in the following pages, describes the single-celled organisms that can give us an idea of what those early forms were like. She goes on to describe the immense range of life in the seas of to-day, which has developed from those simple beginnings—sea urchins and barnacles, corals and clams, lobsters and sea gooseberries, and fish in enormous variety.

At various times in geological history, some marine creatures managed to colonize freshwater rivers and lakes. The shift from one environment to the other can necessitate complex alterations in body chemistry. In the early period the change may not have been too difficult, for then the sea was not very salty. However, as the rocks of the continents continued to be eroded and their minerals washed down to the sea, so the water there became saltier. As a consequence, it became more difficult for creatures to make the move. It is astonishing that some fish today, like the salmon and the eel, manage to make that journey within their individual lifetimes. But they are very exceptional. In general, the freshwater world that Peter Credland writes about is a now separate one, with its own vegetation, its own animal populations and its own particular problems.

The colonization of fresh water turned out to be a crucial stage in the development of life on earth, for it was from freshwater lagoons and pools that periodically dried up, that the first amphibians waddled out onto land. And from them, reptiles, birds, and mammals eventually developed. But curiously, throughout the long progress of evolution, groups of all these land-living air-breathing creatures moved back into water.

Men, too, have learned to live on and around the sea. They have sailed over it and dropped hooks into it since prehistory. They have hung nets in its top layers and dragged trawls over its bed. And so, slowly, they have built up a picture of the immensely complex and infinitely varied patterns of life that it contains.

But in the last few decades, that picture has been wonderfully enlarged. The aqualung was invented. Now, at last, man could follow his mammalian relatives, otters and seals, and plunge beneath the waters, swimming freely, seeing clearly and with sufficient air to remain submerged for long periods. When he did that, he gained a vision of a world that can never fail to fascinate. Between them, Peter Credland and Gillian Standring give a detailed picture of what that world is like, of how the creatures in it live and reproduce, how they communicate and how they are dependent one upon the other. Although our knowledge of the world of water is now great, it is certain that there is still much more to be discovered. For that world covers the greater part of our planet.

David Attenborough.

At Home in the Sea

The course of evolution during the last 2000 million years has brought living things far from their simple origins. We know from fossil evidence that life began in the sea. Even today all plants and animals, including man, have in their cells a liquid—protoplasm—that is very much like sea water in its composition. Man has obviously come far since his distant marine ancestors and he has become perhaps the most successful of land animals. But many other descendants of those same early sea-dwellers have never left the water; instead, they have become adapted to life in the sea, overcoming in a wide variety of ways the recurrent daily problems of feeding, breathing, keeping active, and moving about in the marine environment.

The oceans cover about seven tenths of the earth's surface. But because of the difficulties we face when living and traveling under water, the sea is still relatively unknown and unexplored in comparison with the land. It is hard for us to understand the immense distances and lack of physical barriers in the marine world, where animals move in three dimensions in much the same way as do birds and others that have conquered the air. In fact the two media pose some similar problems, to which animals that live in them have found some similar answers. The sonic wave systems of bats and whales, for instance, are remarkably alike.

The seas and oceans of the world are like the land, too, in that neither of them provides a uniform and unvaried environment. Just as there are few land plants and animals that can live under all possible conditions, so are there few found universally in the seas. During the evolution of each marine organism it has become specialized to live in water of a particular depth, temperature, saltiness, and illumination. Transported to a different part of the ocean, it will at best continue to exist, but be unlikely to grow or

The powerful streamlined body of this 50-foot-long humpback whale is superbly adapted to its life in the oceans. Yet, although whales are the largest and among the speediest of swimmers, they live very differently from most other marine creatures, for they are mammals and therefore cannot breathe without coming to the surface of the water at frequent intervals for air.

Winds and waves have an important effect on marine organisms because surface turbulence helps to dissolve such life-giving gases as oxygen from the atmosphere into the sea. Only the surface layers of the oceans move violently; deep water is always calm.

reproduce; at worst, it will disappear without leaving a trace.

Sea water is a much denser medium than air; a glass of sea water weighs about 800 times as much as a glass filled only with air. Such a dense substance can support far greater weights than can air, and so marine animals do not need strong limbs to stand on. That is one reason why such enormous creatures as whales and giant squids can exist in the sea; the water takes the strain. But because salt water is heavy, it also exerts pressure on the creatures that live in it. Their organs and tissues must be able to withstand crushing pressure from outside. For every 30 feet in depth, the water pressure increases by about 15 pounds to the square inch. Thus at a depth of 300 feet an animal receives the equivalent of 150 pounds' weight on every inch of its body surface. And deep-water animals, which are adapted to the extreme pressure of the abyssal depths, can actually meet their death by "falling upward" with their bodies literally exploding as the outside pressure decreases when they move toward the surface.

Another property of sea water is that it acts as a filter for light and heat. Depending on the turbulence and the amount of mud or sand suspended in it, water absorbs energy waves as they pass through. The red wavelengths of light are removed first, followed gradually by the shorter blue and green wavelengths, and this is the reason for the typical blue-green light that we

see in shallow water. At a depth of 600 feet no light can penetrate even the clearest water. Because heat and light rays are closely associated, the deep oceans are cold as well as dark. Light is less important than heat for animals. But it is essential to green plants, and so plants can live only in the uppermost layers of the sea.

The property that distinguishes sea water from the fresh water of lakes and rivers is the salts dissolved in it. The most important of these is sodium chloride—common salt—which gives sea water its familiar salty taste. But other salts also contribute to the sea's salinity, which varies from place to place. Where large quantities of fresh water enter the sea from rivers and melting ice, the salts are diluted and the salinity is low. In tropical shallows there is rapid evaporation and a high salinity. Enclosed seas that escape the mixing effect of open oceans may also have a high salt concentration. For example, the Red Sea and the Mediterranean, which are enclosed and have high evaporation rates, both average salinities of 40‰, that is 40 parts of salt to 1000 parts of water. The Baltic, on the other hand, is constantly diluted by fresh water and, because it is farther north, has slow evaporation; and so it has a relatively low surface salinity of 7.5‰. The sea's average salinity is 35‰.

The fact that the concentration of salts in sea water is about 35 times as strong as fresh water has an important effect on the functioning of aquatic animals. By a purely physical process called *osmosis*, water passes through a membrane from a dilute solution into a stronger one, provided that the dissolved substance cannot also pass through. Gradually, as the dilute solution decreases and the stronger solution increases in volume, the solutions become equal in concentration. The soft unprotected parts of an animal—its skin, gills, mouth lining, and stomach—act as such a membrane and allow water to pass in or out of its body. The tendency for a marine animal is to lose water into the stronger solution of the environment. Faced, therefore, with the constant threat of dehydration, the sea-dwellers have evolved special kidneys, gills, and certain bodily functions to deal with the problem.

In addition to salts, sea water contains dissolved gases, one of which is oxygen, on which all living things depend for the release of energy from their food. Simple animals, such as sea anemones and worms, can absorb the life-giving gas from water through the surface of their bodies. Other animals live more active lives and need more efficient breathing organs—gills, for instance, or lungs. The waste gas of animals, carbon dioxide, is one of the raw materials on which plants depend for the food-making process of photosynthesis. So both oxygen and carbon dioxide are necessary for the maintenance of life in the sea. The quantities of these gases present in sea water depend on the temperature—warm water retains less gas than cold water does—and on the amount of mixing of water and air. The more turbulent the surface of the sea, the more gas can dissolve in it from the air.

Another source of oxygen for animals is the plants themselves, for they give it out as a by-product of photosynthesis. On land and in shallow water, plants maintain the supply of oxygen, but in the sea the amount of oxygen contributed in this way is small, because plant life is restricted to the upper layers. Thus, most

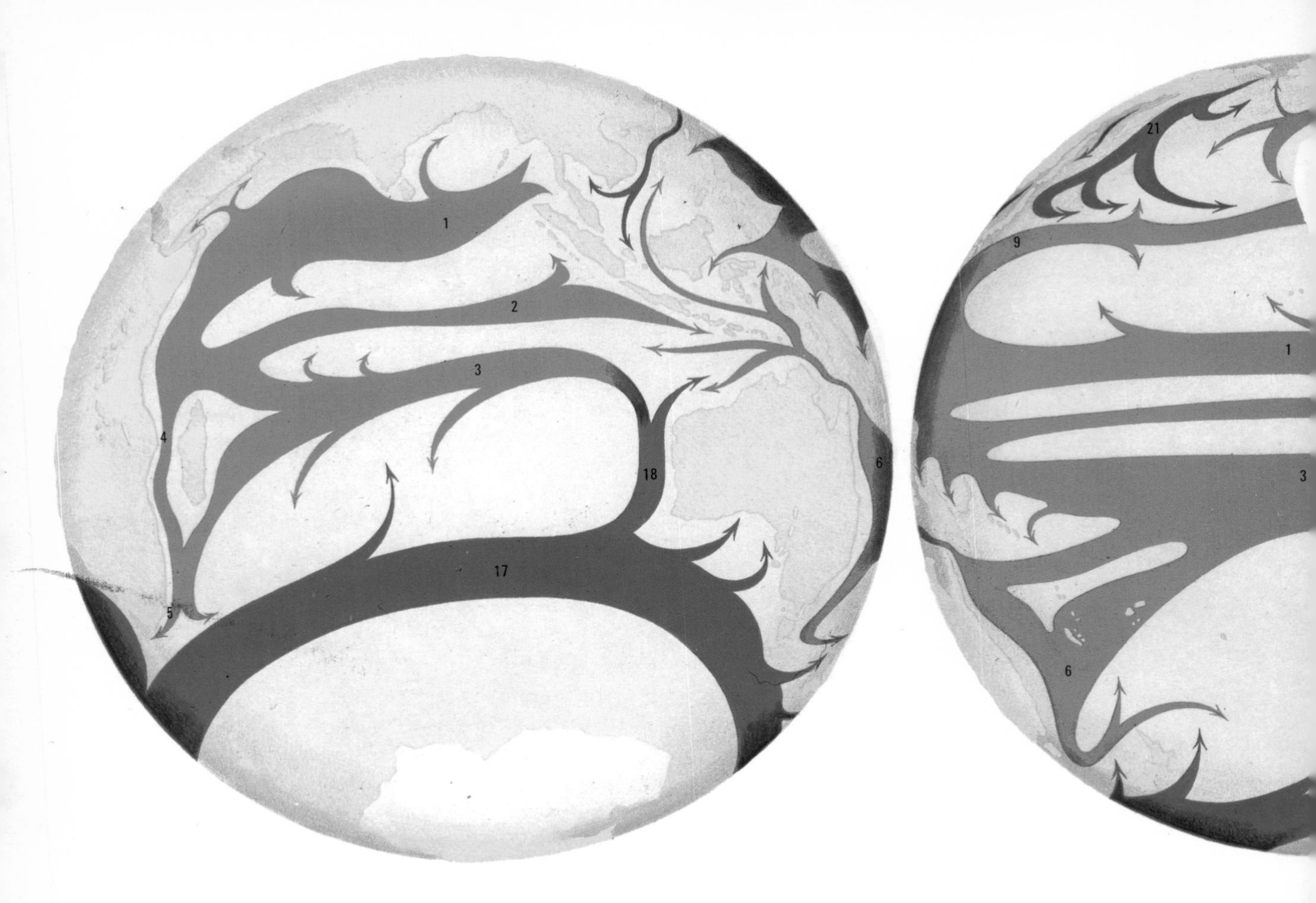

The Major Surface Currents of the World

Warm Currents

1 North Equatorial Current
2 Equatorial Countercurrent
3 South Equatorial Current
4 Mozambique Current
5 Agulhas Current
6 East Australian Current
7 North Pacific Current
8 Alaska Current
9 Kuroshio (Japan Current)
10 Guinea Current
11 Brazil Current
12 North Atlantic Drift
13 West Greenland Current
14 Gulf Stream
15 Florida Current
16 Antilles Current

Cold Currents

17 West Wind Drift
18 West Australian Current
19 Peru (Humboldt) Current
20 California Current
21 Oyashio
22 Falkland Current
23 Benguela Current
24 Canary Current
25 Labrador Current

In all the oceans, the water constantly moves in a series of currents, which are caused by changes in water temperature, winds, and the earth's rotation. Because such currents determine the composition of sea water at various places and depths, they are largely responsible for the distribution of marine plants and animals, and they also affect the climate—and thus the distribution of living things—in the lands whose shores they wash.

marine animals depend on the oxygen that slowly diffuses down to them from the air above.

The chief properties of sea water—its density, and indirectly its composition and illumination—largely depend on its temperature. And because these properties determine where the animals and plants of the ocean live, it is perhaps true to say that temperature is the greatest single influence on the distribution of life in the sea. But although different areas of the ocean may be warmer or colder than others, there are no clearly marked boundaries between them. Often greater variation in temperature occurs at different depths than at different latitudes. For example, around Antarctica, where cold and warm water meet at latitude 50° South, the water is cold—35°F or less—both at the surface and on the seabed, whereas the middle layers are comparatively warm, at about 40°F, having been carried southward by the Equatorial Current. The water of the Atlantic off the west coast of Britain is several degrees warmer than the water of the same ocean at the same latitude off

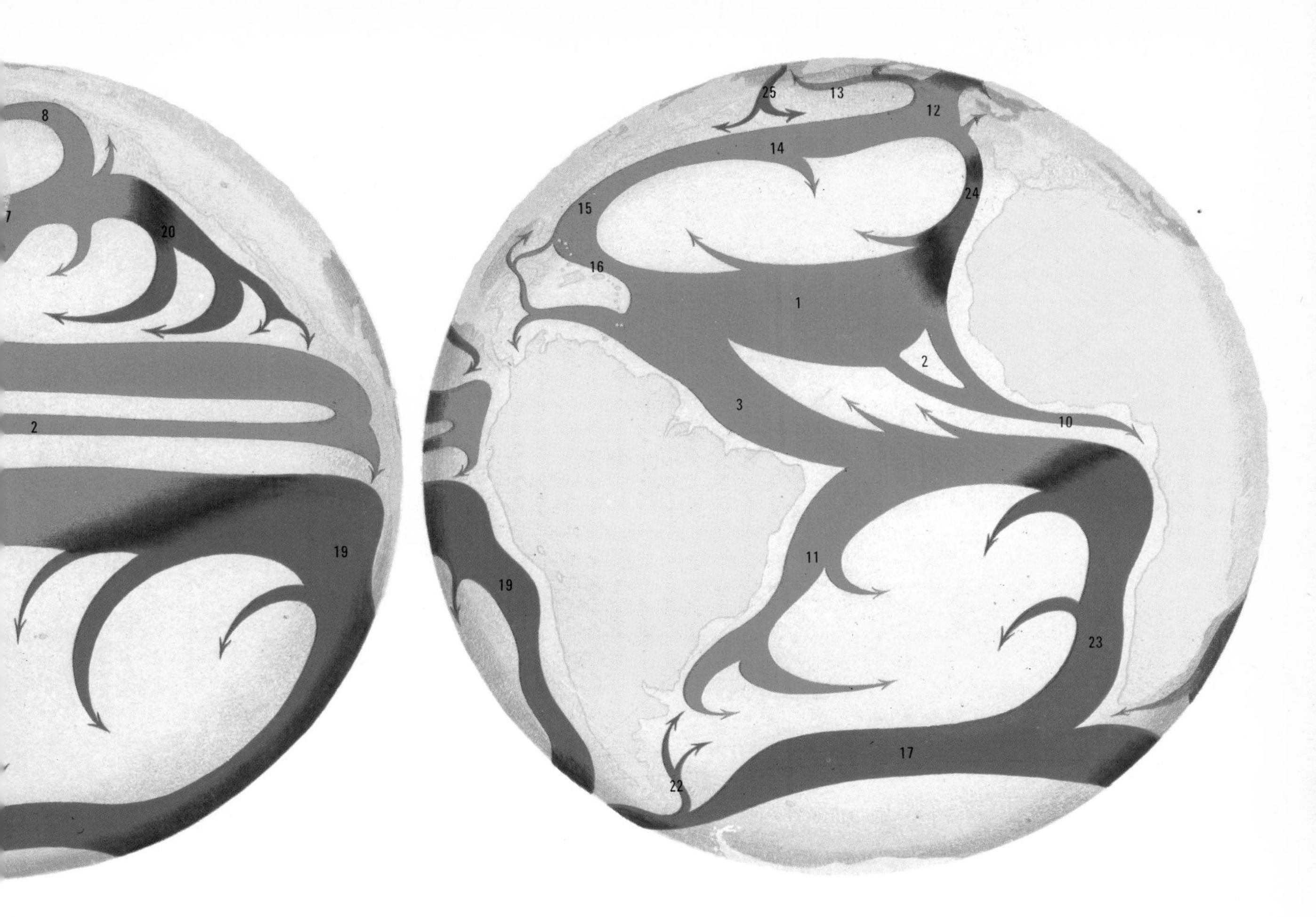

the coast of North America, where there is no Gulf Stream to bring the warm water. The seas cannot, therefore, be divided as sharply into geographical zones as can the land.

Distribution of many plants and animals is, nevertheless, as clearly related to temperature in the sea as on land. Penguins live throughout the Southern Hemisphere, from as far north as the Galápagos Islands near the equator down to the continent of Antarctica in the far south. In every natural habitat of these birds—even at the equator—there is a cold offshore current, which carries plenty of planktonic animals to feed the fish and mollusks on which the penguins live. Contrary to popular belief, then, penguins are not associated necessarily with ice and snow, but with very cold sea water. Ocean currents are caused by a combination of prevailing winds, the earth's rotation, and the varying densities of water at different temperatures. It is these currents, rather than the distance from the equator or the poles, that largely determine the distribution of marine animals.

In the Northern Hemisphere, water warmed by the sun's rays falling on it more or less perpendicularly at the equator travels northward in a series of currents moving in a clockwise direction. On reaching the cold regions, the water loses some heat, becomes denser, and sinks, forcing less dense water to the surface. While near the seabed, this surfacing water has picked up many mineral salts from the decaying remains of plants and animals. Other currents at the surface now carry the cool but fertile water southward to the main fishing grounds of the north Atlantic and north Pacific before returning to the region of the equator. A similar pattern occurs in the Southern Hemisphere but here, because of the effect of the earth's rotation, the currents flow in a counterclockwise direction.

In all the world's great oceans a constant circulation of water takes place. The biggest movements are close to the shorelines of the continents, and the central areas—the "deserts of the sea"—remain relatively static and unmixed. Replenishment of nutrient minerals in sea water is

Shallow tropical waters, such as those of the Pacific around Tahiti (above), are warm and well lighted but poor in minerals. Thus, although a greater variety of marine animals lives here than in the cold waters of the Antarctic Ocean (right), the mineral-rich polar seas are able to support larger numbers of the few animal species that are adapted to life in these harsh conditions.

essential to the life of marine plants and animals. So the areas where the richest population of marine organisms occurs are the areas where warm water mixes with cold, more fertile water.

At the surface of the sea the interaction between winds and water creates waves. This often violent movement of the upper layers plays an important part in increasing the gases dissolved in sea water. It also produces some very difficult living conditions for animals and plants on the shores. Out at sea, most animals are more or less unaffected by storms, because the effect of strong winds and waves is hardly felt beneath the surface layers of water. Some ocean birds, such as the albatross, even benefit from strong winds, because their gliding flight depends on the movement of air between wave peaks.

As well as temperature and the movements caused by currents and winds, another important factor that influences the distribution of plants and animals in the sea is the depth of the water. Not surprisingly, adaptations for various types of habitat and ways of life can be most clearly seen on the vertical scale. To live in shallow water, where illumination is good, temperature relatively high, and water pressure low, an organism requires a different structure and a different pattern of behavior from those of an organism that lives in the depths.

Projecting around most of the continents and larger islands of the world are areas known as *continental shelves*, which slope gently away from the shoreline to an average depth of 700 feet. The width of the shelf varies considerably, from almost nothing to as much as 800 miles; then suddenly the gradient of the seabed becomes steeper, forming the *continental slopes*, which are the highest and steepest escarpments on earth.

These precipitous slopes are the true boundaries of the continents.

Beyond them stretches the mid-ocean floor, at an average depth of nearly two miles, broken by underwater mountains, trenches, and plateaus. And the seabed, like the earth's crust lying above sea level, is liable to violent movements of various kinds—volcanic eruptions, earthquakes, and rifts along faults, which can disrupt life in the ocean as drastically as they do on land.

So much, then, for the general picture of the varying conditions that affect marine organisms. The ability of animals and plants to survive in a range of conditions varies enormously. Some animals, such as barnacles on rocks and parrot fish among corals, are highly adapted to their restricted habitats. Others—for instance, the sperm whale and the herring gull—are less specialized and therefore quite widespread in the oceans of the world.

In the following pages we shall look at life in the oceans and on the fringes of the sea. First of all we shall see how and what marine animals eat, because finding food and avoiding being eaten are basic daily tasks in most animals' lives. We shall then examine some of the interesting underwater partnerships that have evolved, with one creature living more or less permanently with another, sometimes gaining and sometimes losing in the process. Whatever its relations with other organisms, an animal must have ways of identifying and communicating with them. Toward such ends a number of special forms and behavior patterns have evolved, and we shall take a close look at these, followed by a study of how animals move in the sea.

Finally, it will improve our understanding of the immensely varied saltwater world if we take a look at some examples of marine environments. How, for example, does a balanced community of organisms exist at the poles, at the tropics, or in habitats at the edge of the sea, where land and water meet? From our consideration of such communities it is but a small step to a consideration of questions that trouble people today—questions concerning the effects of the activities of man. The menace of pollution and the hope of conservation are topics closely linked with the basic principles of ecology on which our entire study of marine life rests.

To Eat or be Eaten

Life in the sea, no less than on land, depends ultimately on the sun for its existence. In both environments green plants use energy from sunlight to combine simple raw materials—water, carbon dioxide gas, and inorganic salts—making carbohydrates and proteins. This process, called *photosynthesis*, has oxygen as an important by-product. Some animals consume plants in order to build their own tissues. These plant-eaters may, in turn, be consumed by meat-eaters, including scavengers. All these animals breathe in the oxygen released by plants and in return give out carbon dioxide to help replenish the supply used by plants. Dead tissues and waste products of both plants and animals are broken down into simple reusable substances by fungi, decay bacteria, and a host of other small organisms.

In the oceans plant life is restricted to shallow water and the surface layers. Life-giving sunlight cannot penetrate far in sea water, so photosynthesis by green plants cannot take place below about 600 feet in clear water, or 150 feet in turbid water. Another limiting factor for marine plants is the availability of carbon dioxide. Most of the carbon dioxide in the sea is absorbed from the atmosphere and not produced by marine animals. Thus the plants are dependent on the steady absorption of carbon dioxide at the surface.

Two groups of plants—the seaweeds and the tiny floating plants that make up the *phytoplankton*—produce directly or indirectly all the food for the animals living in the oceans, in spite of the plants' limited distribution. Enormous amounts of plant food are eaten by the primary consumers, the herbivores. These animals, in smaller numbers, are the food of the secondary consumers, carnivores, and so on up the food pyramid. At the top are the few species of large predators on which nothing normally preys. Thus the complex food web of the sea is built up.

It has been calculated that 100,000 planktonic plants will feed 10,000 tiny planktonic crustaceans, supplying enough food for 1000 herring smelt or 100 adult mackerel, which might be a meal for 10 codfish or a single porpoise. So a tenfold reduction in the number of living things supported occurs in each step up.

Seaweeds are eaten by browsing mollusks,

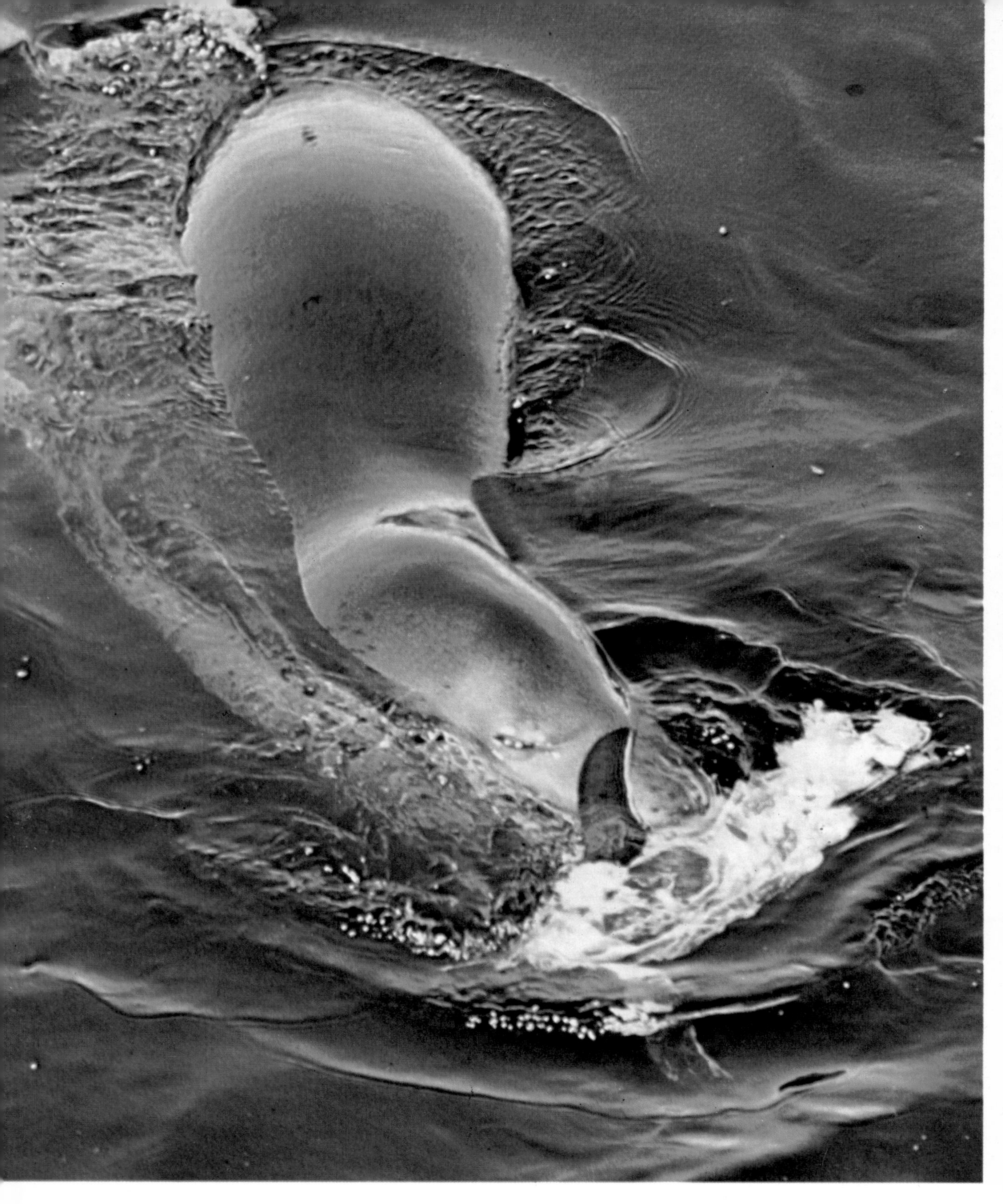

Some animals are specialized plant-eaters, such as the flat periwinkle feeding on bladder wrack (left). These mollusks continually browse on the seaweed as they move slowly about. Other animals, such as this leopard seal devouring a penguin after chasing it under the water (above), are voracious predators that must hunt and kill other animals in order to live.

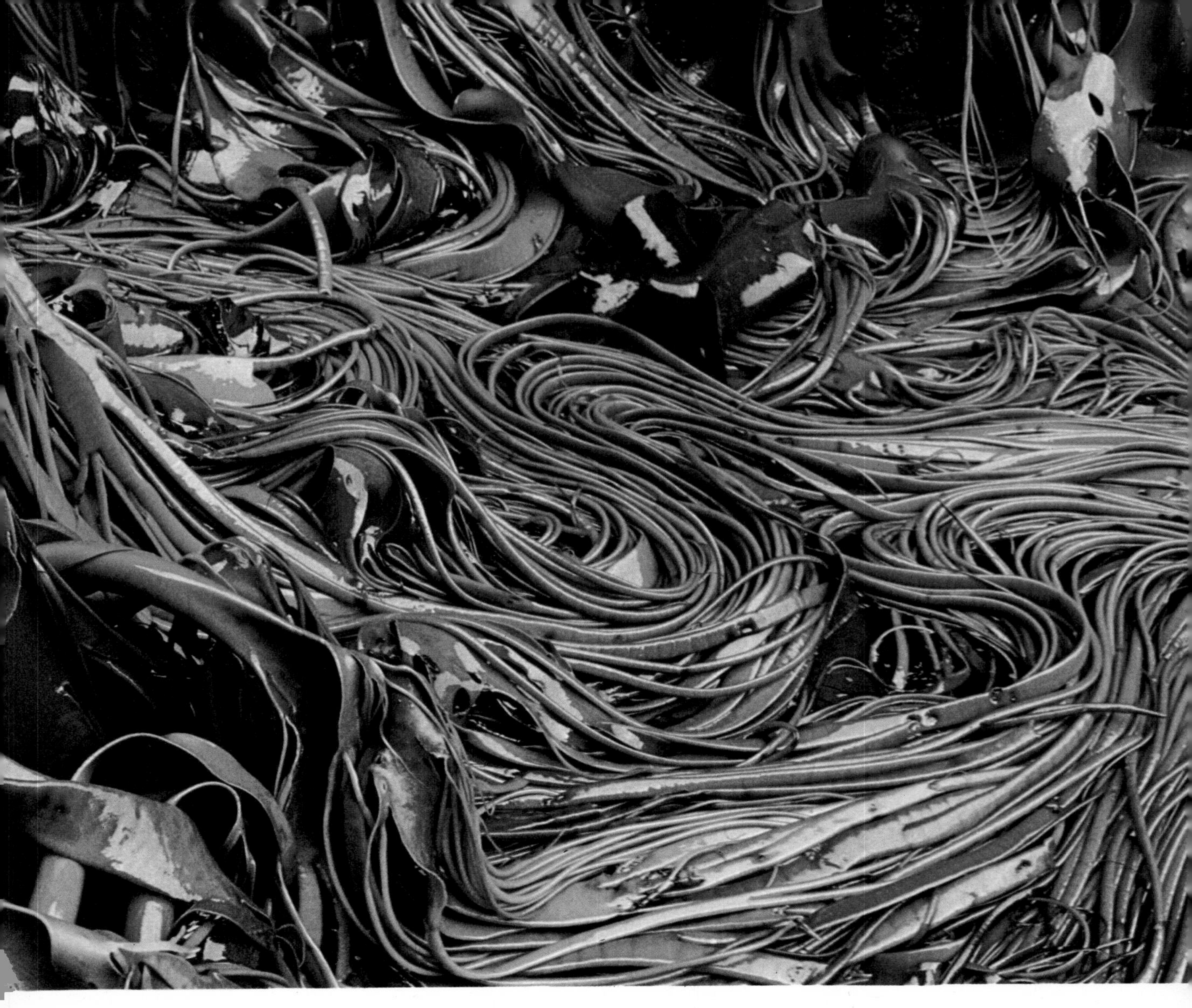

which are the food of predatory invertebrates, including starfish and octopuses. *Demersal*, or "bottom-living," animals such as these are eaten by demersal fish: skate, dogfish, halibut, and others that often end up as human food. The larger decaying remains of dead animals fall to the sea floor as debris and feed scavenging lobsters and flatfish. The smaller remains are removed by detritus-eaters and decay bacteria, which form a layer of slime over rocks and sand particles. Like the rest, the decomposers finally return to the sea as dissolved minerals.

Even from this outline it is clear that the food web of the sea is very complicated, with each organism contributing to and dependent on many others. It is important to realize that, like a piece of knitting, if one of the threads is broken the whole complex web begins to unravel.

We see a classic example of this in the anchovy fishing industry of Peru. As a result of modern fishing methods, the huge schools of anchovies feeding in the plankton-rich waters of the cold Peru (Humboldt) Current are being greatly reduced each year. The flocks of seabirds, whose droppings form guano, feed mainly on anchovies and are in serious danger of starvation because they cannot find enough food. The precious supply of guano, a valuable agricultural fertilizer, is soon used up but only slowly replaced. When the anchovy schools are further reduced by the periodic occurrence of "El Nino," a warm current, lacking in plankton, that flows over the cold current at the surface, the natural balance is even more upset.

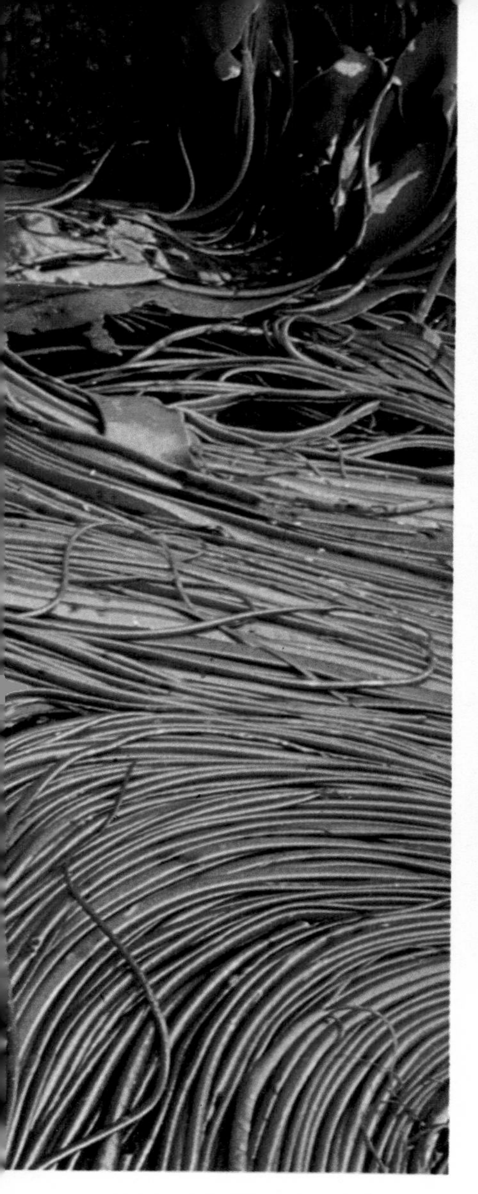

Around the rocky shores of the world's oceans grow such seaweeds as bull kelp (left) and serrated wrack (below), the diet of countless numbers of animals that live among them. The blue-rayed limpet (above) is using its rasplike tongue on a tough stalk of tangleweed, from which it has eaten out a deep scar. If it were not for the limpet and other herbivorous mollusks, the continuously growing weeds would be much more abundant on our shores because there are few vertebrates that feed on them.

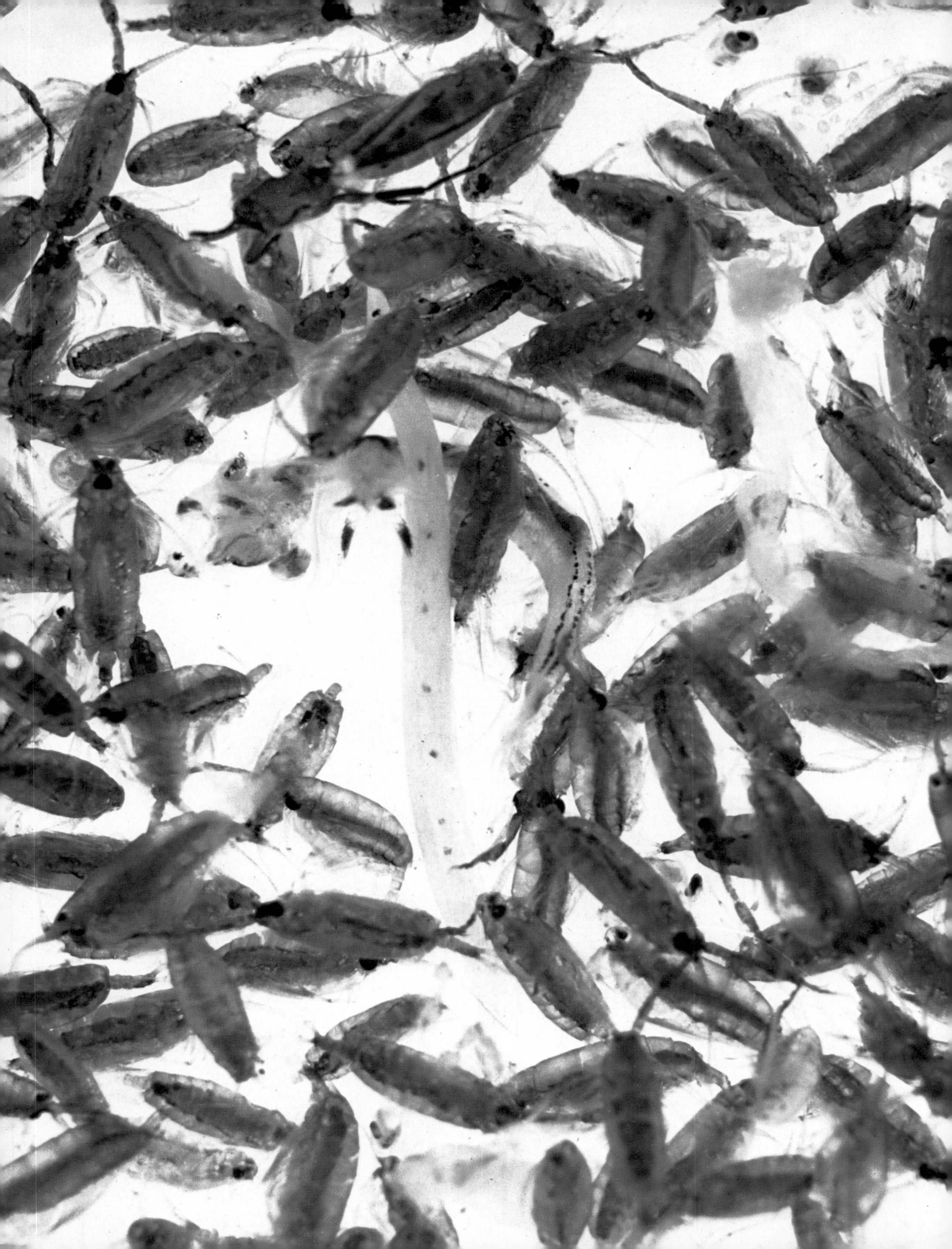

Above: the often observed phosphorescence of the sea's surface at night is at least partly due to the presence in the plankton of Noctiluca, *a minute plant that emits flashes of blue-green light when the water is disturbed by waves and the movement of ships.*

Animals need ready-made food to provide energy and materials for their cells. A basic supply of food that is available all the year round is seaweed. Around the margins of the seas, mainly where there are rocky shores, masses of seaweeds grow, corresponding to the shrubs and trees on which land herbivores browse. Seaweeds are all algae. From the whole plant kingdom only the algae are adapted for survival in the sea. They are able to cope with such rigorous conditions as continuous pounding by breakers, regular drying out at low tide and soaking at high tide, and the predations of large concentrations of browsing animals.

The variety among seaweeds is reflected in their colors—brown, red, and blue-green, depending on the combination of chlorophyll and other pigments in their cells. Some of the most widespread are the large brown seaweeds commonly known as "wracks." Bladder wrack, knotted wrack, thongweed, and others cover rocks in the intertidal zone. At the lowest tides another brown seaweed—oarweed or tangleweed—is exposed as an undulating mass of straplike fronds.

In the uppermost layers of the sea live the many floating animals that make up the zooplankton. These are some of the tiny crustaceans, worms, and larvae—here, of course, much magnified—to be found in a sample of tropical water.

In nearby rock pools there are smaller kinds of seaweed. Thin flat sheets of bright green sea lettuce and purple lavers, *Enteromorpha* (which looks like green hair), dark red, curly "Irish moss," and feathery, pink *Corallina* are all common on north Atlantic shores. And there are many more.

Unlike land plants, seaweeds do not undergo seasonal changes. Instead there is a constant year-long cycle of growth, death, and replacement, and so they provide a continuous source of food for the mollusks that browse on them. If it were not for the multitudes of limpets found on most rocky shores, the weeds would be even more abundant. Like all herbivores, limpets need a hard grinding surface to help them procure their food, and instead of teeth they have a long filelike tongue, or *radula*, which works back and forth to rasp the weeds off the rocks. Each limpet

The great basking shark is one of the largest fish, sometimes reaching a length of more than 40 feet; yet it feeds entirely on microscopic planktonic organisms. As it swims along with its mouth open, plankton-rich water flows in and is filtered out through filaments lining the gills. The suspended food is strained off and then carried into the shark's stomach to be digested.

moves slowly about its territory, clearing it of vegetation before moving on.

Periwinkles of various species also feed on seaweeds, each kind preferring a different food plant. Flat periwinkles are usually found among bladder wrack, their smooth shells varying in color from bright yellow through olive green to black and orange stripes. Higher up the shore, rough periwinkles with ribbed shells browse on the thongweed. Above the high-water mark, in the *splash zone*, the small periwinkle feeds on lichens and so is almost a land animal. And in the tide pools and the area just below low tide live less familiar browsing mollusks, such as the sea hares, which feed on soft seaweeds.

In the open sea a second great basic supply of food is easily found: vast swarms of plankton float in the upper layers of the sea, and the planktonic plants are comparable to the low-growing "carpet" plants on land—grasses, mosses, and lichens—on which grazing ruminants feed.

The microscopic plants of the phytoplankton are, like the seaweeds, all algae, and are the only plants adapted to a permanent free-floating marine life. They exist in enormous numbers in many areas of ocean, but because of their size it is not possible to see individual plants with the naked eye. However, if you tow a fine-mesh net slowly behind a boat for a few minutes and look at its contents under a microscope, you will see the amazing variety of shapes and colors displayed by these single-celled plants.

Above: these mature herrings, enormous schools of which inhabit cool northern oceans, also feed largely on plankton. Their staple food is the tiny planktonic shrimp Calanus *(below, highly magnified), and the herring themselves are preyed upon by all sorts of carnivorous fish, whales, porpoises, and seabirds. The herring is also an important and popular food fish of Northern Europe.*

Most numerous are the *diatoms*, each like a tiny transparent pillbox containing brown, green, or yellow pigments similar to chlorophyll. Their shapes vary from flat plates to long ribbons and slim pencils.

After the diatoms, the commonest plants of the plankton are the *dinoflagellates*, which are even more varied in shape. Their name comes from the Greek word *dino* meaning "spinning" and from the hairlike *flagella* that each carries and by which it can propel itself through the water. Dinoflagellates have red "eye spots" that are sensitive to light, and many, although not all, contain chlorophyll. Those without chlorophyll feed on other planktonic organisms and are on the borderline between plant and animal.

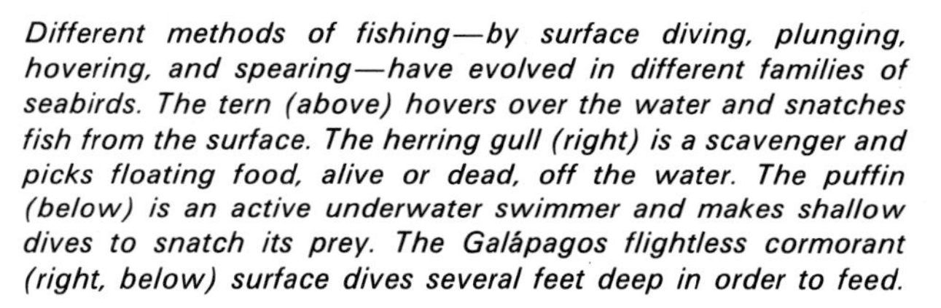

Different methods of fishing—by surface diving, plunging, hovering, and spearing—have evolved in different families of seabirds. The tern (above) hovers over the water and snatches fish from the surface. The herring gull (right) is a scavenger and picks floating food, alive or dead, off the water. The puffin (below) is an active underwater swimmer and makes shallow dives to snatch its prey. The Galápagos flightless cormorant (right, below) surface dives several feet deep in order to feed.

Anyone who has been on a boat at night will be aware of the luminous light on the waves, especially in the ship's wake. This is due to the presence in the plankton of *Noctiluca* ("night light"), a dinoflagellate that releases flashes of brilliant blue-green light. When the water is disturbed it seems to sparkle like a fireworks display as a result of the activity of these tiny organisms. The function of this light is still a mystery but it may be simply a way of getting rid of extra energy produced by the cell.

The animals of the plankton belong to many different groups. On the whole they are larger than the planktonic plants and can be seen under a hand lens. Tiny larval forms of jellyfish, which swim with a pulsating motion; darting translucent arrow-worms; spinning larval stages of worms, winkles, barnacles, crabs, and starfish; and bristling full-grown crustaceans only a fraction of an inch long—all are important members of the zooplankton along with the eggs and young of many kinds of fish. During the warm months between 4000 and 5000 animals, and 75,000 diatoms and dinoflagellates, may live in a cubic yard of sea water near the surface.

This rich mixture of floating plants and animals is the food of many fish, including the great basking shark. Like a grazing animal, this plankton-eater has no need to chase its food, but it must be able to take in large amounts quickly. To do so it uses a filtering system. As it swims slowly along at the surface with mouth wide open, water passes in and is sieved through finely divided rakers along the gills. The water leaves through the gill openings, while the suspended plankton is strained off and carried into the shark's stomach.

Filter-feeding requires little energy and in plankton-rich waters even large mammals can support themselves in this way. The blue whale, at 120 tons probably the largest animal that has ever lived, has fibrous curtains of *baleen*, or whalebone, hanging down inside its mouth. Through these sieves masses of tiny planktonic shrimps called *krill* are filtered from the water drawn in by the expansion of the whale's pleated throat. Fifty years ago, hundreds of these giants could be seen "blowing" as they fed on the great schools of krill in the Antarctic Ocean, but they have been brought to the edge of extinction by the whaling industry.

Another plankton-eater, the herring, is an extremely important animal of northern seas.

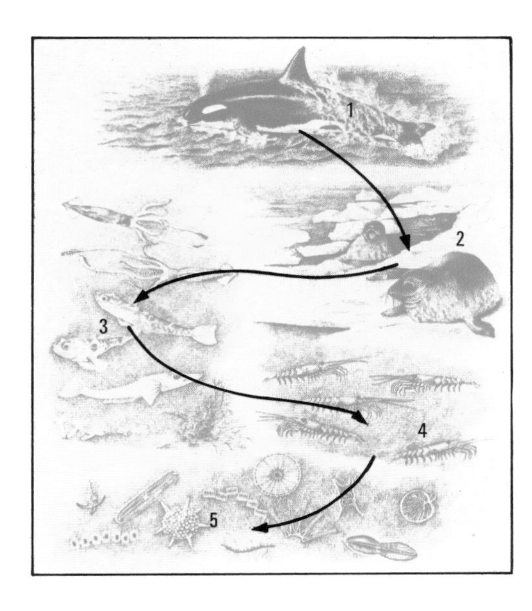

Killer Whale Food Chain

1 Killer whale
2 Weddell seal
3 Antarctic fish
4 Krill
5 Plant plankton

A favorite food of the voracious killer whale is seals, and the seals in turn prey largely on squid and fish, as shown by the arrows at left. The food of these animals is extremely varied, but it includes the tiny shrimps known as krill, *and the krill feed on the microscopic planktonic plants that we call* diatoms *and* dinoflagellates. *Thus the killer whale is at the top of one food pyramid, in which each layer has roughly 10 times as many organisms as the layer above it.*

Mature herrings spawn in coastal waters in the spring or fall. The millions of eggs sink to the seabed, where many are eaten by demersal fish such as haddock, attracted annually to the herring spawning grounds. After about 10 days, the herring larvae hatch and, still with some yolk attached, they swim up to the surface waters and become part of the plankton. The tiny elongated larvae begin to feed on the phytoplankton when they have used up all the yolk. They are themselves eaten in large numbers by other planktonic animals, particularly arrow-worms and sea gooseberries. The surviving larvae become carnivorous and grow rapidly on a diet of tiny crustaceans. At about one and a half inches long the fish's shape changes suddenly to that of an adult, covered with scales. Its main food for the rest of its life is the tiny shrimp *Calanus*. Adult herrings gather in schools over the continental shelf area, sometimes covering up to 100 square miles. Thousands of these fish are eaten by other *pelagic* (open sea) fish such as cod, sharks, and rays, and by fin whales, rorquals, killer whales, and porpoises, which often follow the schools.

Herrings and other pelagic fish provide food for a variety of seabirds. Although they belong to several families, seabirds are all alike in that they feed at sea. However, there are many ways of catching fish, and the different feeding methods used by seabirds are a good example of what is known as *adaptive radiation*—in other words, ways in which basically similar animals have developed distinct specializations.

The method of fishing used by some birds is to fly low over the sea and snatch up food from the surface. Inshore, the terns are expert at this. These graceful "sea swallows" hover above the water with heads down, then drop vertically to snatch up sand eels, minnows, or shrimps in their pointed beaks. Having caught a fish, the tern sometimes performs a noisy "fish flight" before offering it to its mate, but usually the fish is quickly swallowed head first.

The larger and clumsier herring gull is also a surface-feeder, but it will eat any kind of food, dead or alive, floating at the surface. Gulls are buoyant swimmers but rarely dive. Noisy crowds of them follow fishing boats to pick up the fish offal, and they also scavenge inland on rubbish dumps, because they can walk as easily as they can swim. Out at sea, surface-feeding birds are mainly albatrosses, petrels, and shearwaters, which return to land only to nest. These birds are known as "tubenoses" because of the long tubular nostrils on each side of their hooked bills.

Another group of birds dives from the surface and feeds under water. In the Northern Hemisphere it is the members of the auk family—puffins, guillemots, and razorbills—that feed like this, and they closely parallel the habits and ways of life of the penguins of the Southern Hemisphere. Auks sometimes fly far out to sea but their short wings are better adapted for paddling than for flying and they can remain submerged for half a minute.

One of the most spectacular seabirds is the goose-sized gannet. Like its relatives in the tropics, the pelicans, boobies, and tropic birds, it

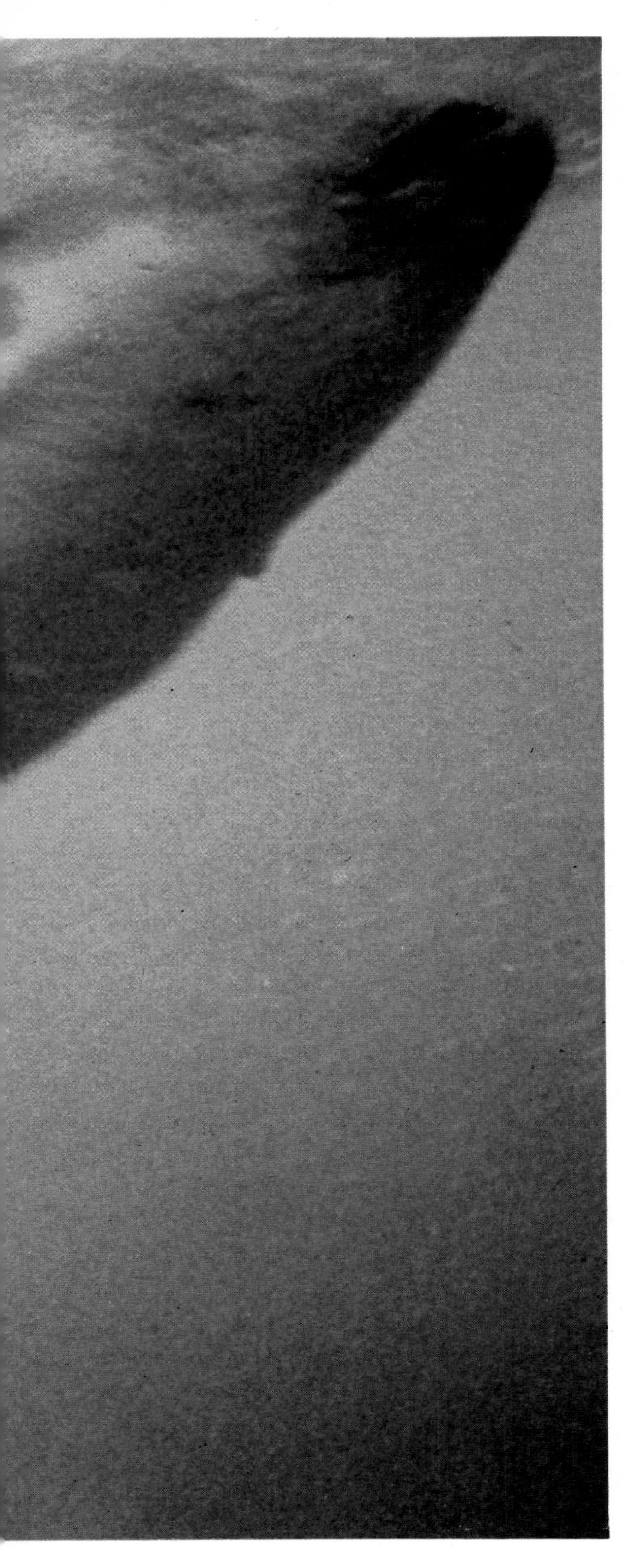

catches its food by plunging into the water from a height. Gannets often dive from a height of 100 feet above the surface of the water, but they stay under the surface for only a few seconds. They seize herrings, mackerel, whiting, or small haddock in the open bill, and usually swallow them under water.

Seabirds that stay on the water surface or fly with fish in their beaks are in danger of attack by "pirate" birds. Skuas in the temperate and polar latitudes and man-of-war birds in the tropics are aggressive marauders. They harass other birds until food is disgorged or dropped. Skuas have rarely been seen to catch live fish, but they eat eggs and nestlings of other birds as well as offal and carrion.

Beneath the surface, as above it, there are hungry predators that feed on all kinds of marine animals. Many cartilaginous fish are bottom-feeders but the great white shark feeds on pelagic fish and other sharks. These masters of streamlining also eat seals, whales, turtles, sea-birds, and offal. Their notoriety as man-eaters is usually exaggerated, but they are attracted by blood and will seize any injured or dying animal. Sharks are armed with razor-sharp teeth, each species having its own particular tooth shape. Like those in other predatory fish the teeth point inward and the arrangement of the teeth around the wide curved mouth makes the jaws ideal weapons for grasping and tearing their prey.

Another voracious hunter with a bad reputation is the grampus or killer whale. One of the toothed whales, the killer whale has up to 50 large pointed ivory teeth in its jaws. Killer whales hunt in packs. They prefer warm-blooded animals as food and hunt seals, sea lions, walruses, and penguins in water or on ice. They also attack dolphins, porpoises, and large whalebone whales. After the kill they leave much of their prey to be finished off by sharks and other scavengers.

Seals, in turn, are themselves predators. The leopard seal of the Antarctic, for instance, lies in wait among the pack ice for its favorite prey—penguins. These it skins with one quick toss and then it swallows them whole, head first. It is one

The 20-foot-long great white shark is a man-eater, but it normally preys on fish, including other sharks. An inhabitant of tropical coastal waters, it can tear great chunks of flesh out of any available food with its saw-edged razor-sharp teeth. The shape of the teeth is peculiar to this species of shark.

Not all predators hunt actively. A Portuguese man-of-war (left) is a floating colony of polyps suspended from a gas-filled sac, which sting and paralyze passing fish. The octopus (above) lurks under rocks, its eight arms ready to seize slow-moving crabs, mollusks, and bottom-living fish.

of the most specialized as well as one of the largest of the seals. It has exceptionally long and sharp teeth and its long sinuous body gives it extra speed and strength and enables it to catch penguins rather than fish—the usual diet of seals.

The richness of the sea's food supply makes it unnecessary for all predators to be fast-moving and active hunters. Many are more or less sedentary animals, waiting for their food to come to them. The octopus rarely uses its eight "arms" for crawling. It hides in rock crevices—well disguised by its rocklike coloring—and grabs fish, crabs, and mollusks with the rows of suckers on its arms. Contrary to popular belief, these suckers do much less damage than the octopus's strong, beaklike jaws. Stories of people being seized by the long arms are probably due to the animal's great curiosity. For an invertebrate the octopus is remarkably intelligent, and will stretch out its arms to investigate any moving object that might prove to be a meal.

On the rocks among which the octopus lurks

live many kinds of sea anemones. In spite of their name and appearance these are not plants, but simple animals, with a bag of double-layered cells for a body, surrounding an all-purpose body cavity that holds sea water. The graceful waving tentacles around the mouth carry batteries of stinging cells, some of which entrap any animal that swims into them while others inject a paralyzing poison into the victim. Although they are sedentary throughout their adult lives, sea anemones can catch and eat quite large fish.

Jellyfish belong to the same group of simple animals as the sea anemones. They also catch their food by means of stinging cells in their trailing tentacles, as they float or swim gently through the water. Although most jellyfish cause no more than an unpleasant irritating sting to humans, their poison is strong enough for their feeding purposes. And a few of them, such as the "lion's mane" and "sea wasps," produce some of the most powerful venoms known in the animal kingdom.

Sometimes a number of small, harmless animals live together in a powerful colony that behaves as one unit, in which each individual has its own specialized function of feeding, defense, or reproduction. Such a colony is the "Portuguese man-of-war," a relative of the sea anemones

Scavengers, such as the colorful Pacific lobster (above) and the edible sea urchin (below), eat the dead remains of plants and animals. The lobster tears big pieces of debris apart with its powerful pincers; the sea urchin has five hard teeth with which to break up its food, which includes shells of dead animals.

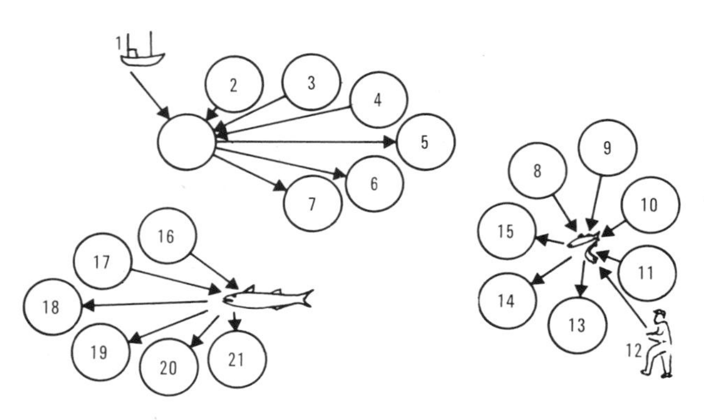

The Feeding Cycle of the Mackerel

January-June
(Plankton feeding)

1 Midwater trawling
2 Hake
3 Mackerel shark
4 Common dolphin
5 Copepod *(Calanus)*
6 Copepod *(Anomalocera)*
7 Small fish

July-September
(Inshore feeding)

8 Spiny dogfish
9 Gray seal
10 Hake
11 Cod
12 Angler
13 Herring
14 Sprat
15 Sand eel

October-December
(Bottom feeding)

16 Spiny dogfish
17 Cod
18 Polychaete worm
19 Amphipod
20 Opossum shrimp
21 Shrimp

How one carnivore eats and is eaten. Mackerel are the vital link in different food webs throughout the year: from January through June, when they feed near the surface; from July through September, when they come inshore; and during the winter, when they feed at the bottom. Arrows on the key point from *mackerel to their prey, and* toward *them from their predators.*

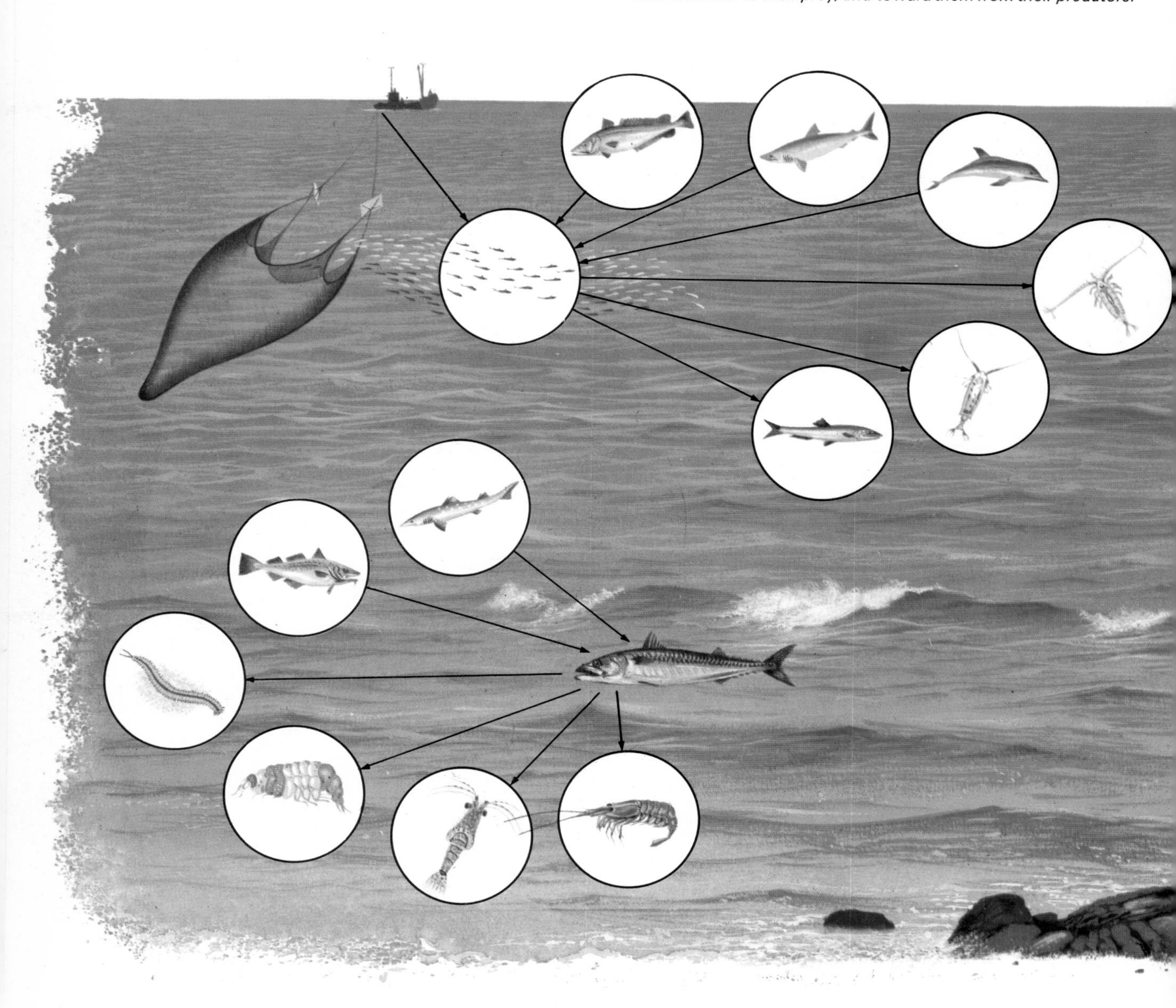

and jellyfish. Its beautiful scientific name of *Physalia* belies the dangerous trailing tentacles, which can paralyze and hoist up to the feeding animals fish as big as full-grown mackerel.

It is not always necessary to kill, however, in order to eat. Among the most essential animals in the sea, as on the land, are those that feed on dead and decaying plants and animals—the scavengers. The death of all kinds of organisms provides a varied diet for the lobster. This large crustacean sometimes uses its powerful pincers to seize living prey, but it usually eats the bigger pieces of debris that fall to the seabed. The lobster has complex mouthparts, versatile tools with which it selects and macerates its food before the mill-like stomach grinds it up.

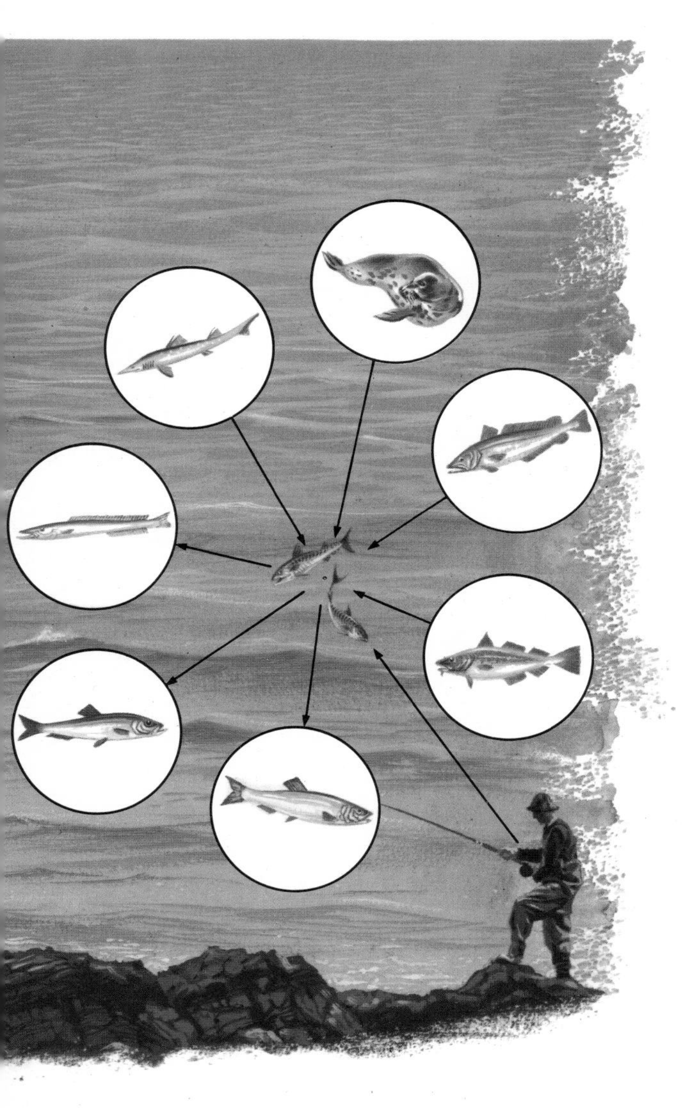

The sea urchins are another group of scavengers. Some, such as the heart urchin and "sand dollar," burrow into sand and feed on the decaying matter that coats the sand particles. Others are more active, gliding slowly over sand and rocks on tiny suction feet or movable spines. Inside its delicately colored shell, the edible sea urchin has a remarkable apparatus for obtaining its food: its suctorial mouth is in the center of the underside of its body, and inside the mouth there are five hard teeth that move up and down in V-shaped frames. With this structure, called "Aristotle's lantern" after the famous Greek who first described it, the urchin is able to break up the hard parts of dead animals, such as the bones and shells. Sea urchins have even been known to chew their way through steel pier supports.

The mechanism of filter-feeding found in plankton-eaters can also be used to remove tiny suspended particles of decaying organic matter from sea water. Detritus-eaters using this method are found among rocks and in places where the seabed is sandy or muddy. They use a system of siphons to draw detritus-rich water from near the seabed into their bodies, where it is filtered by the gills. Clams, oysters, scallops, and mussels are bivalve mollusks that feed in this way. Their feeding places differ, however: the clam pulls itself down into the sand by a muscular foot; the sedentary oyster and the free-swimming scallop lie on the bottom rather than burrowing; and the mussel binds itself to rocks by tough threads. Unfortunately, these mollusks are not able to select the type of food they filter from the water and are sometimes poisoned by sewage poured out over them from coastal towns.

Many kinds of worm, including the ragworm, live in U-shaped burrows in sand or mud, sucking in a current of water and particles by the action of flat paddles along their bodies. The food particles are trapped in a sticky mucous net, which, at intervals, is rolled up, eaten, and renewed by the worm. Fan worms and peacock worms display their colorful crowns of tentacles above the sand and use them to sweep the water for food.

In shallow waters and in the upper layers of the open sea, food is varied and usually abundant. In the ocean deeps, however, where no plants can grow because of the lack of light and where animals occur singly rather than in schools or colonies, finding food is more of a problem.

In the absence of plant life all the deep-sea animals are carnivores or scavengers. A constant rain of decaying plant and animal remains falls from above upon the seabed. The crustaceans, mollusks, and starfish are detritus-eaters, like their relatives in shallow water; and they are in turn eaten by fish such as the rattails, which have especially strengthened snouts, mouths underneath, and long, tapering tails that help incline the body at an angle to the sea floor. This posture enables them to root about in the ooze for their prey.

Because the number of animals living at great depths is quite small and the fish are sparse, living prey is hard to come by. Many deep-sea fish have enormous mouths and expanding stomachs so that they can take in and swallow any animal that comes along. They must be able to seize every opportunity of food. The gulper eels have grotesquely shaped heads, with jaws that can open wide enough to engulf fish larger than themselves. To digest such a colossal meal, the balloonlike stomach of a gulper eel can expand to a point where it presses so hard on the heart and gills that it pushes them out of their normal position. Both angler fish and gulper eels have marvelously economical body structures, with poorly developed skeletons, weak muscles, and small brains, gills, and kidneys. This means that they can exist on infrequent meals and do not use much energy hunting for food. Nor do they need to swim about very much: slight movements are enough to maintain their position in the water.

Biologically speaking, there is no such thing as a free meal ticket. This account of the food and feeding methods of a variety of marine organisms shows how true this statement is. Whether it feeds on plants, or animals, or their remains, every creature inevitably makes its own contribution to the food web of which it is a part, both during its life and after its death.

Below: sedentary scavengers, like such mobile ones as lobsters and scallops, also do their bit to keep the oceans clean by eating up suspended particles of decaying organic matter. These fan worms, which live in the sand encased in tubes made of bits of shell and sand particles, extend their feathery tentacles above the sand and use them for sweeping specks of food out of the water.

Fish that live in the depths must be able to eat any food they can find, because the number of abyssal animals is extremely small. The rattail (above) can incline its body at an angle to the seabed and root about in the ooze for its prey, and the black swallower (seen below gulping down a hatchet fish) has enormous jaws and an expandable stomach. Both these deep-sea dwellers can survive on very infrequent meals.

Living Together in the Sea

Survival in the sea does not depend solely on the availability of food, however. Just as important, and one of the most fascinating aspects of animal ecology, is the existence of partnerships between animals, where the association is closer than just living in the same habitat. There are several degrees of interdependence. Sometimes two different species regularly live together although it is hard for us to imagine what they gain from the relationship. This has been called *commensalism* or *mutualism* and a good example can be seen in the snapping shrimp and the goby, a crustacean and a warm-water fish that often share the same sand burrow. The shrimp digs the burrow and keeps it clean, and the fish apparently comes and goes as it pleases without contributing to the housekeeping.

It is not always easy to draw a clear distinction between such a relationship and the kind in which one or both partners derive some benefit. An anemone living on a whelk shell occupied by a hermit crab gets transport from the crab, of course. But what is the reward for the crab—unless, as often happens, the two animals share their food?

Sometimes both the animals involved benefit in very obvious ways and the association is then called *symbiosis*. Mullet, or goatfish, living among coral reefs indicate that they need the attention of a "cleaner" by standing on their heads. A small striped wrasse then carefully goes over the larger fish, removing parasites and infected tissues from the scales, fins, and mouth. Here the mutual advantages are obvious, for parasites kill as many fish as predators do, and the wrasse, advertising itself by bright colors and using a regular corner of the reef as a "cleaning station," gains its food in this way.

If a "cleaner" fish or shrimp occasionally takes a bite out of a fish it is attending, and also lives fairly permanently with it, it is not difficult to see how the parasitic relationship might have arisen. There is not a great difference between the way of

In the sea, as in other environments, survival often depends on cooperation between animals of different kinds. Close relationships—some examples of which are shown here—may be mutually rewarding, but often only one partner benefits.

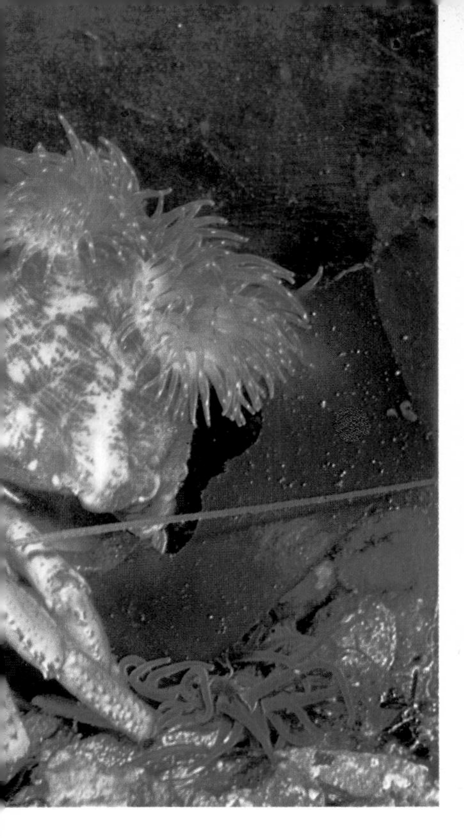

Below: it is unlikely that bull sharks profit from the remoras that attach themselves to their hosts by head suckers, thus gaining transport and bits of left-over food. But a hermit crab in a whelk shell (left) may get food and camouflage in return for transporting clinging sea anemones, and the sweetlips (above), a resident of coral reefs, is cleansed of parasites by small wrasse seen emerging from its mouth.

Camouflage is the chief means of protection for many otherwise defenseless marine animals. In a matter of seconds, plaice can change their coloring to match a background of pebbles (above) or fine gravel (below) by redistributing colored pigments contained in special cells in the fish's skin. The rapid change is an automatic response to the immediate environment.

life of a remora, clinging harmlessly to a shark and feeding on the shark's leftovers, and that of a lamprey, which uses its suckerlike mouth to rasp away at the tissues of its host fish. A parasite is completely dependent for food on its host, which in turn suffers from the association. The parasite may even kill its host in the end, but not before it has produced large numbers of larval parasites. Scars from attacks by lampreys are often found on basking sharks and whales and on freshwater fish such as trout, so we know that even the lamprey's attack is not always fatal.

Smaller parasites are more common than large ones and often more dangerous. *Sacculina*, a relative of the barnacles, begins life as a tiny planktonic larva that attaches itself to the leg of a crab, burrows through the shell into the bloodstream, and infests the crab's body. One of the side effects of this parasite is to castrate male crabs. Later, the adult parasite appears as a flat growth beneath the abdomen of the crab, which soon dies. Before this happens, however, thousands of young *Sacculina* larvae will have been released into the sea.

Because parasitic animals lead very specialized lives, they often look quite unlike their free-living relatives. To look at a "whale louse" you would not guess that it, too, is a kind of barnacle. It can be up to 10 inches long and is attached to its host by a horny three-pronged head through which it feeds on the whale's blubber. Another kind of whale louse clings by many hooked legs to the skin of a right whale or a humpback whale, but seems to do relatively little harm. Most slow-swimming whales carry large numbers of barnacles as well as these true parasites but the barnacles are sedentary plankton-eaters and are just as likely to be found growing on rocks, ships, or wooden piles as on living animals.

Because there is continuous competition between predator and prey—whether we consider free-living organisms or parasites and their hosts—it is not surprising that many kinds of defense have evolved among animals. Like land animals, marine creatures can defend themselves by hiding, by fighting, or by taking flight. Although some fish hide themselves physically—such as the blenny, which takes refuge in large empty shells, or the eel, which buries itself in the sand—many rely on being difficult to see in the open.

Flatfish, such as plaice and flounders, can make themselves invisible by matching exactly the color of the sand on which they lie. When

The pipefish disguises itself by imitating the eelgrass among which it lives. It can sway like the plant stems as it swims vertically, or it can even stand motionless. Staying quite still is often an essential element in good camouflage.

such a fish moves to a different patch of sand the color of its skin can become lighter or darker in seconds by means of the *chromatophores*. These are cells in the fish's skin containing black, brown, and yellow pigments. In combination with other cells containing pearly white guanine they can produce solid colors or mottled patterns as the pigments automatically disperse or concentrate in response to whatever the fish sees below it. The camouflage is further improved by the fish's habit of flipping sand over its body with its fins. However, camouflage is not simply a matter of matching the color or pattern of the back-

A contrasting defense mechanism to camouflage is conspicuous coloring. The alarming appearance of this dragon fish warns potential enemies to keep their distance. If the warning goes unheeded and the predator continues to approach, the fish can inflict a painful, or even fatal, wound by means of the venomous tips of its dorsal spines, but its flamboyant colors are a first line of defense.

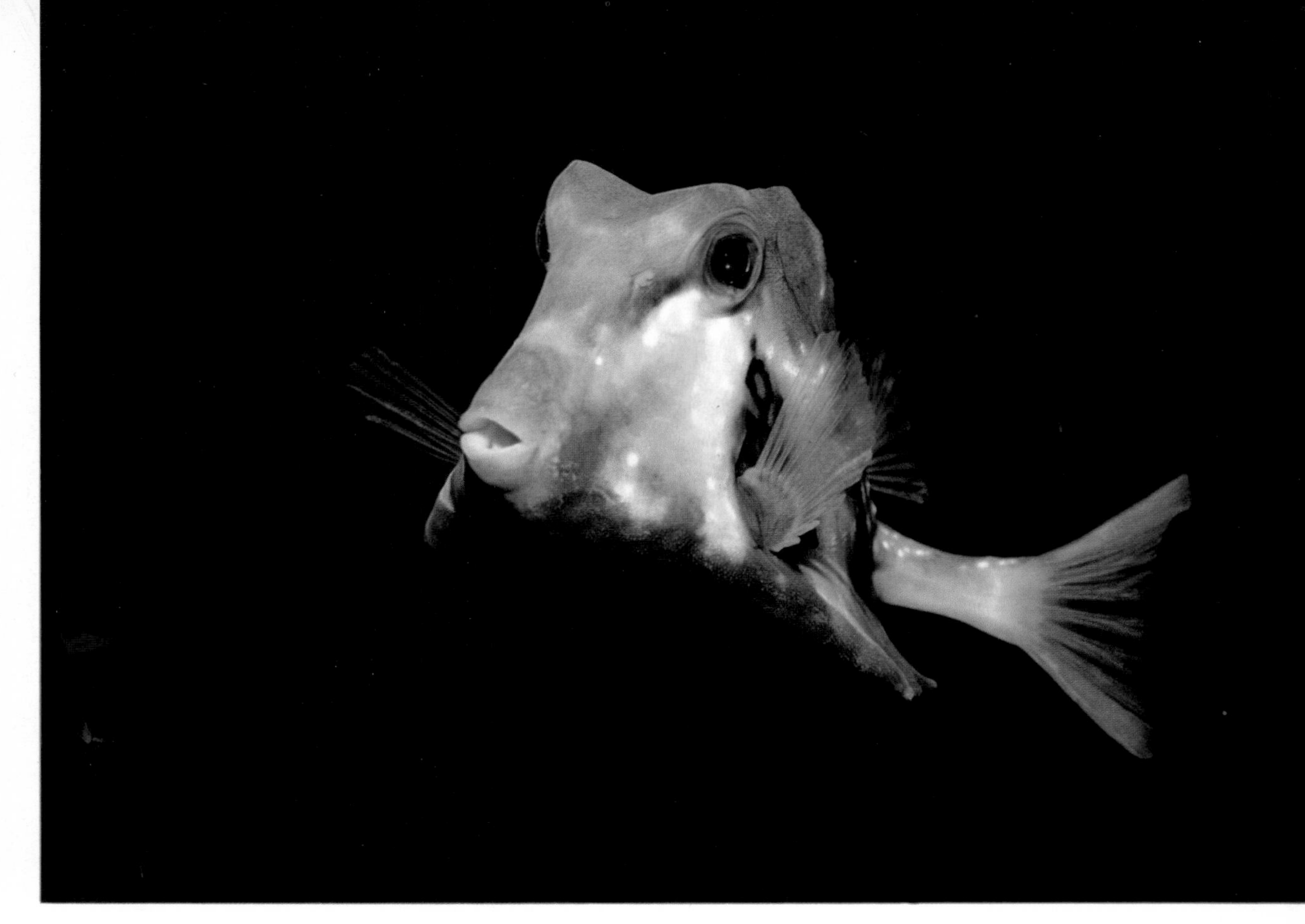

Above: because the trunkfish is encased in armor-plating made from fused and thickened scales, from which only its eyes, mouth, and fins protrude, it cannot bend its body to swim with its tail as most fish do, and so it rows itself along with its paired frontal fins. As an added security measure, some species are conspicuously colored or have poisonous mucus covering their bodies.

ground; telltale shadows must also be hidden. A flatfish hides its shadow by lying on it.

Many small crustaceans, such as shrimps and prawns, living in shallow water above a sandy seabed, are very difficult to see because their bodies are transparent. Their tough shells are sometimes like clear glass, or they may have scattered spots of color. In either case the animal is well camouflaged until it moves.

Nearly all fish and many mammals living in well-lit open waters are camouflaged by countershading. The mackerel conceals itself from predators in this way. Seen from above, its back has dark and light blue stripes that blend with the water's dark ripples, helping to conceal it from hunting seabirds. Its silver-white underside, seen from below, matches the brightly lit surface, and is equally effective against predators beneath. Viewed under water from the side, the mackerel's body appears as one color, that of the water.

In contrast, cardinal fish have various patterns of conspicuous black and white stripes and spots that break up their body outline and make them difficult to recognize as fish. Most animals with zebra stripes show examples of this disruptive coloration. Some butterfly fish have bright false "eyes" near the tail or on the gill covers to mislead an enemy as to the direction in which the intended prey will flee. Such vivid patterns are normally found among tropical fish living in shallow, clear water, where visibility is good.

Conspicuous coloring may also be effective in defense. The dragon fish, a relative of the scorpion fish and stonefish, has long winglike fins and gaudy stripes of black, red, and white. Its fantastic shape serves as a warning to discourage other animals from coming too near it. But if a predator is not put off by the alarming appearance of the dragon fish and seizes it, it receives a painful or even fatal wound from the powerful venom injected by the spines along the dragon fish's back.

Effective camouflage can also be achieved by imitating part of the habitat. Pipefish swim vertically among tall eelgrass, their bodies swaying like plants. A seaweed-encrusted "rock" that suddenly moves could be a well-disguised scorpion fish, a stonefish, or some kind of angler fish. Sargassum fish mimic their habitat to the extent of having fins with knobs and tassels exactly like the bladder-covered weed among which they live.

Many crabs have shells whose shapes, colors, or texture blend in well with the rocks and weeds of their habitat. Others use parts of the surroundings to disguise themselves. The decorator crab attaches pieces of seaweeds, sponges, and seamats to its own shell to make itself virtually invisible. And the sponge crab cuts out a cap of sponge with its claws and holds it over its back like an umbrella.

Protective disguise is only one of the many kinds of defense made use of by marine animals, however. In fact, most kinds of defense adopted by men through the ages are paralleled by those of marine organisms. Coral animals live in impregnable "castles" that they build in fantastic shapes from minerals they extract from sea water. Armor is beautifully shown by the crabs

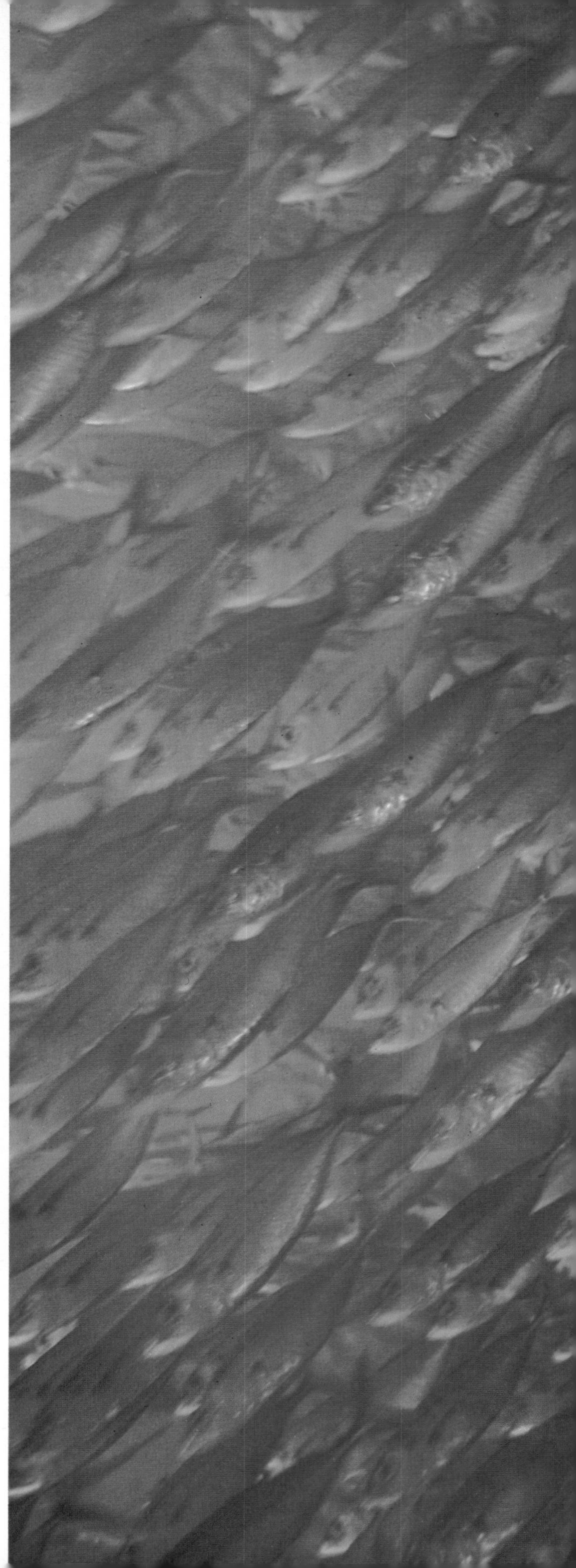

Whether in solitude on the bottom or as members of free-swimming schools, all fish have evolved ways to protect themselves from predators. The eyed electric ray (above) can produce electric discharges of up to 40 volts from the "batteries" in its broad fins. Shad (right) seek safety in numbers by collecting in schools of hundreds of thousands; this gives individual fish a much better chance to survive.

and their relatives, whose hard external skeleton is jointed and hinged to allow movement. A crab must shed its protective garb at intervals so that the animal inside can grow. At such times it is as vulnerable as a medieval knight without his armor. The trunkfish, whose scales are thickened and fused together to form a complete box-like shell, swims about like a tiny army tank, paddling its rigid body slowly along in the water with its front fins.

The squids and cuttlefish, which are relatives of the octopus, vanish from their enemies behind "smoke screens": each animal has a bag of ink that it squirts into the water to cover its retreat from danger.

Other defensive weapons that protect fish from the many predators of the oceans include the 40-volt shock produced by the huge flat "wings" of the electric ray. The specialized muscles in these expanded front fins form a "battery" and can be used offensively to stun prey or animals that menace the sluggish but well-camouflaged ray lying on the seabed. Its close relative, the stingray, has a row of sharp poison spines along its whiplike tail. Curved spines can be sprung out like switchblades from the tail of the surgeonfish. The triggerfish gets its name from two or three needlesharp spines that can be raised and locked into position from the fin along its back: these can be used as weapons or to secure the fish in its hiding place, in a rock crevice. The puffer, or porcupine fish, has a body like an inflatable pincushion, which protects it from all but the most aggressive attackers. Most fish that "fight back" in one of the ways described have conspicuous warning colors.

For a fish with no built-in weapons against predators a communal life provides at least some protection. When they live together in a large school, each individual is, so to speak, "hidden in the crowd" and has a much better chance of avoiding an enemy than if it were alone. Herring, mackerel, anchovies, and other school fish swim together in thousands.

Fish in schools are governed by two contrary stimuli. On the one hand they are drawn together by the attraction of their kin; on the other, they respond to a repelling stimulus, which allows

The battle is not only between predator and prey. In coral reefs fish of the same species often fight for possession of territory. Distinctive coloration, shown here by the black and white false moorish idol, is one way of proclaiming territorial rights.

each fish sufficient space in which to swim and breathe freely. Fear of predators will overcome this second stimulus and allow them to pack together more tightly. There is also a remarkable form of communication in a school, with every fish changing direction simultaneously.

Another advantage to the concentration of many fish into one small area is that there is less chance of them being located in the ocean. However, hunters such as whales and seabirds have their own ways of dealing with this problem. Birds have excellent eyesight and when one gull or gannet plunges to a school it has found, other birds over a wide area see and join it. As for whales, they navigate by high-frequency echolocation and can detect schools over a mile away.

Besides the conflict between predator and prey, there is also continuous competition between members of the same species for living space, food, and mates. In the 4000 kinds of school fish this competition is less important than the need to protect the species as a whole. But many kinds of small fish living in and around coral reefs constantly do battle with their own kind and with such violence that they are oblivious to external danger while so enraged. For example, species such as damselfish are strongly territorial in behavior and even young ones, only a fraction of an inch long, will fight fiercely for their own feeding and breeding areas. This aggressive territorial behavior is a way of spacing out individuals in the habitat. Most territorial aggression takes the form of threat displays and sparring rather than serious fighting. Once established in its own territory a fish is normally able to drive away all rivals.

A strange territory is occupied by the little clownfish, which is brilliantly patterned in orange and white. It has a unique immunity to the stings of certain sea anemones, possibly through the use of a protective mucus produced by the anemone itself. The clownfish can live safely among the anemone's tentacles, well-protected against foes. It defends its chosen home valiantly against rival clownfish, to whom its flaglike colors announce its "ownership."

Bright colors that normally act to keep fish well spaced out must change at times to allow mating to take place between two fish that would otherwise be hostile. For example, boldly marked black and white damselfish turn a dull gray while mating and spawning. It is quite common, too, for young fish to be more highly colored than sexually mature adults. The only time two fish with the same gaudy colors live harmoniously together is in a permanent mated pair, such as occurs with the blue angelfish or the Beau Gregory (a species of damselfish). These conjugal pairs chase other fish from their territory even more fiercely than a single fish does.

In the sea, as on land, the male must first establish his territory by showing himself in it and driving away intruders. He must then attract a female and stimulate her to mate. Some male fish use their colors to do this. In the breeding season the cuckoo wrasse of the north Atlantic, whose normal color is a subdued pattern of blues and greens, takes on a rainbow effect emphasized by a dramatic white patch over the head. He shows off his new colors in a spectacular courtship display to the plain pinkish brown female. A similar flamboyance marks the courtship behavior of the male dragonet, a bottom-living fish of shallow coastal waters, who fans out his brilliant fins before the nearest female. When she responds by moving toward him, he supports her on his broadly spread pelvic fins as they swim together up to the surface, where the eggs are laid and fertilized.

In shallow, well-lit water or within a close-knit school, finding a mate is not difficult. However, the main problem for fish in deep water, apart from finding food, is making contact with a suitable partner. As with all the essential functions of living organisms, various ways of coping with this have evolved. Deep-sea fish often have distinctive patterns of luminous organs on their bodies, making them easy to identify in the dimness. Some, such as the lantern fish, even have a different pattern in males and females.

The problems of finding mates are reduced in fish that establish a permanent relationship once they have found each other. The most striking example of this can be seen among the angler fish. When a young male angler, helped by his sensitive nose, finds a female, he sinks his teeth into her head and hangs on. The skin of his jaws becomes joined to that of the female, most of his organs degenerate, and the two bloodstreams become continuous. The male, less than one third the size of the female, thus becomes parasitic on

The little clownfish inhabits a unique territory among the tentacles of a sea anemone, to whose painful stings it is immune. Perhaps the gaudily striped fish pays for its lodging by attracting other small fish into the death trap that constantly awaits them.

his mate, and he remains dependent on her for the rest of his life. The female, consequently, is guaranteed fertilization of her eggs.

Another way of overcoming the problem of scarcity of mates is to include both male and female organs in the same animal, which is then called a *hermaphrodite*. True parasites in situations where only one can be supported by a host are often hermaphrodites, but such bisexuality is rare among free-living vertebrates. Still, it does exist—for instance, in some sea perches, and in the small deep-sea fish known as *Benthalbella*.

A number of marine mollusks, such as the sea hares and sea slugs, are also hermaphrodites. Common sea hares mate in chains, with each

Courtship behavior is often an important prelude to mating. Above: after the male dragonet has stimulated the female by fanning out his brightly colored fins, the nuptial pair join and swim side by side up to the surface, where they shed their eggs and sperm. In a similar transformation, the male cuckoo wrasse (right, with his smaller mate in the background) brightens the north Atlantic breeding season by acquiring brilliant colors enhanced by a white head patch; he then parades his newfound beauty before the female, whose color is always pinkish brown.

animal acting as a male to the one in front and as a female to the one behind. The chain may close to form a circle, or the leading animal may play only the female role while the last one plays only the male role. Mollusks in general show several unusual reproductive methods. The octopus, squid, and cuttlefish are mollusks with separate sexes, and the males have a special arm that is spoon-shaped at the tip, to carry sperms to the female. Sometimes this special tip breaks off during the transfer, and when this was first observed it was thought to be a parasitic worm in the body of the female.

And once mated, what of the offspring? In the ocean, growing up is a hazardous business. The sea itself and the hungry predators take an enormous toll of the young produced by jellyfish, worms, crustaceans, mollusks, fish, and mammals alike. It is small wonder that marine animals (with the exception of the mammals, where parental care is highly developed) must produce large numbers of fertilized eggs to ensure the survival of a few. The sea hare is known to produce over 4 million eggs during the summer. This animal must be one of the most prolific alive, but few of the eggs survive all hazards and reach maturity. Some animals, such as the female herring, can lay tens of thousands of eggs a year. Other fish, for instance the angler fish, lay even more prodigious numbers, in the range of 1

The gaping mouths of these female angler fish distinguish them clearly from the much smaller males, which have pincerlike mouths. Once the male has found a mate, he secures the relationship by sinking his teeth into her and becoming a permanent parasite, totally dependent on her for the rest of his life. The female provides food and oxygen in return for fertilization of all her eggs.

million. We have seen that these eggs and the young animals that hatch from them are an important part of the plankton, which is at the base of all marine food webs.

Only animals that provide some form of protection for their growing young can afford not to produce huge numbers, which is wasteful of both energy and cell materials in the female. Paradoxically, the cartilaginous sharks and rays, which are less highly evolved structurally, provide for their young much better than the more advanced bony fish. Sharks have internal fertilization, the male using special claspers to put sperms into the female's body. Therefore, each egg is almost certain to be fertilized. In bony fish the sperms are shed rather haphazardly over the eggs in the water, and of course fertilization is more chancy, with a resultant high wastage.

Some sharks—the dogfish, for instance—and the skates and rays lay their fertilized eggs, supplied with yolk for food, enclosed in tough horny shells in which the embryos develop. Such shells, called "mermaids' purses," have coiled tendrils that become entangled in seaweed. The young fish take up to nine months to hatch. In many sharks, though, the eggs are never released, and instead the embryos develop for as long as two years inside the body of the female. When finally set free, they are active young fish, well able to fend for themselves. This means that a shark needs to produce relatively few offspring. A blue shark, for example, may give birth to only about 30 young during its lifetime.

In the sea horses and related pipefish, it is the

male who looks after the young. The female deposits her eggs in a brood pouch on the underside of the male's body. After fertilization the eggs are retained in the pouch, nourished by special substances secreted by the male. After a few weeks the fully formed babies emerge to cling to surrounding weeds, like miniature versions of their parents.

The female octopus is a very devoted mother. After internal fertilization, she lays her eggs in groups like tiny bunches of grapes, which she sticks to the roof of her den under the rocks. For several weeks she guards the eggs continuously, at intervals squirting a current from her siphon to aerate and clean them. During this time she does not eat at all, and octopuses have been known to starve to death while caring for their eggs. Octopuses, too, hatch as miniature adults, not as larvae like most mollusks. Young octopuses are free-swimming members of the plankton, but later they sink to the seabed and become much less mobile. The argonaut, or paper nautilus, which is related to the octopus and is found in warm seas, provides an unusual "nursery" for its young: the female has two special arms that secrete a thin papery shell and hold it around her body. Inside this shell the eggs are laid and cared for.

Large crustaceans also show a good deal of parental care. The female lobster carries her fertilized eggs wherever she goes until the young hatch. The eggs have a sticky outer covering that

Sea hares (shell-less mollusks that browse on soft seaweeds below the low-tide mark) are hermaphrodites, with both male and female organs. To reproduce they exchange sperms and lay millions of eggs, but only a few of them survive and reach maturity.

Above: the reason why most fish lay very large quantities of eggs is that only a few can reach maturity, because little or no parental care is provided. The butterfish, a rather eel-like fish of coastal waters, does give some protection, however: it coils itself around its mass of eggs, which are often laid in shallow pools on rocky or muddy shores. The eggs are guarded until they hatch.

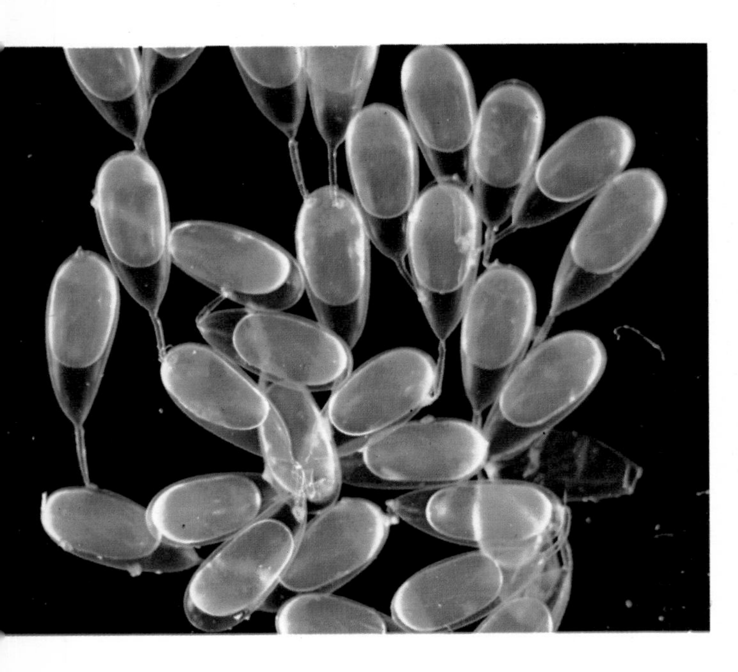

The female squid attaches her eggs—which have been fertilized inside her body—to surrounding rocks in strings, each egg protectively encased in its own transparent capsule.

glues them to the mother's hairy legs. In one kind of mantis shrimp the female collects the glutinous mass of eggs after they are laid and carries them in her large clawlike pincers. In another species the egg mass forms a cape over the back of the female as she squats in her burrow. Marine relatives of the terrestrial wood lice, such as the wood-borers, have a brood pouch of special overlapping plates on the underside of the female, with a built-in aeration system for the developing eggs. The majority of young crustaceans leave the mother as soon as they hatch, either as larvae or as miniature adults, but those of the shrimp-like *Arcturus* are carried about by the female, clinging to her very long antennae like washing pegged to a line.

Thus animals in the sea, as on land, live in various kinds of relationship with one another. Two of the most basic relationships, those of predator and prey, and parent and offspring, are at the extremes of a range that includes many degrees of dependency; as we have seen, no animal can live its life in total isolation.

Each of the fertilized eggs of sharks, skates, and rays contains a supply of yolk for food and is well protected by a tough, horny shell about three inches long. The coiled tendrils on the ends of this "mermaid's purse" become entangled in seaweed, and so the egg is anchored down while the young fish develops inside. These fish are among the few in which internal fertilization takes place.

Contact and Communication

All animals must be able to identify potential food, enemies, and mates. Through its senses every animal receives information about its surroundings. It then reacts to this information by various patterns of behavior controlled by the nervous system.

We, as highly evolved terrestrial animals, rely mainly on our eyes and ears to find our way around, with the senses of touch, taste, and smell playing a minor role in our lives. Living in the sea, however, requires a different sensory system because air and water do not transfer information in the same way. Smell and touch become more important in water, and hearing often involves high-frequency receptors.

Many of the simpler marine animals do not possess sense organs in the form of eyes, ears, and noses, but they are still very much aware of changes in their surroundings. A sea anemone, for example, quickly withdraws its tentacles and pulls its body into a shapeless blob when the tide goes out, or if the animal is suddenly buffeted by water, shaded, or prodded. The anemone's body is covered with a network of nerve cells, some of which sense what is happening around the animal while others make it react accordingly. Yet the anemone, like the related coral and jellyfish, has no spinal cord, brain, or sense organs.

The density of water makes it a better medium than air for transmitting the small, rhythmic pressure changes caused by vibrations. Thus many marine animals, including fish, have an added sense that enables them to detect and respond to vibrations caused by moving objects in the water around them. On most fish there is a visible line of special scales running along each side of the body. This line marks the position of the lateral line sense organ—a small canal that runs just below the surface of the skin and opens to the outside by tiny pores. The canal is filled with fluid, into which hairlike sensory cells protrude. Any small pressure changes or movements

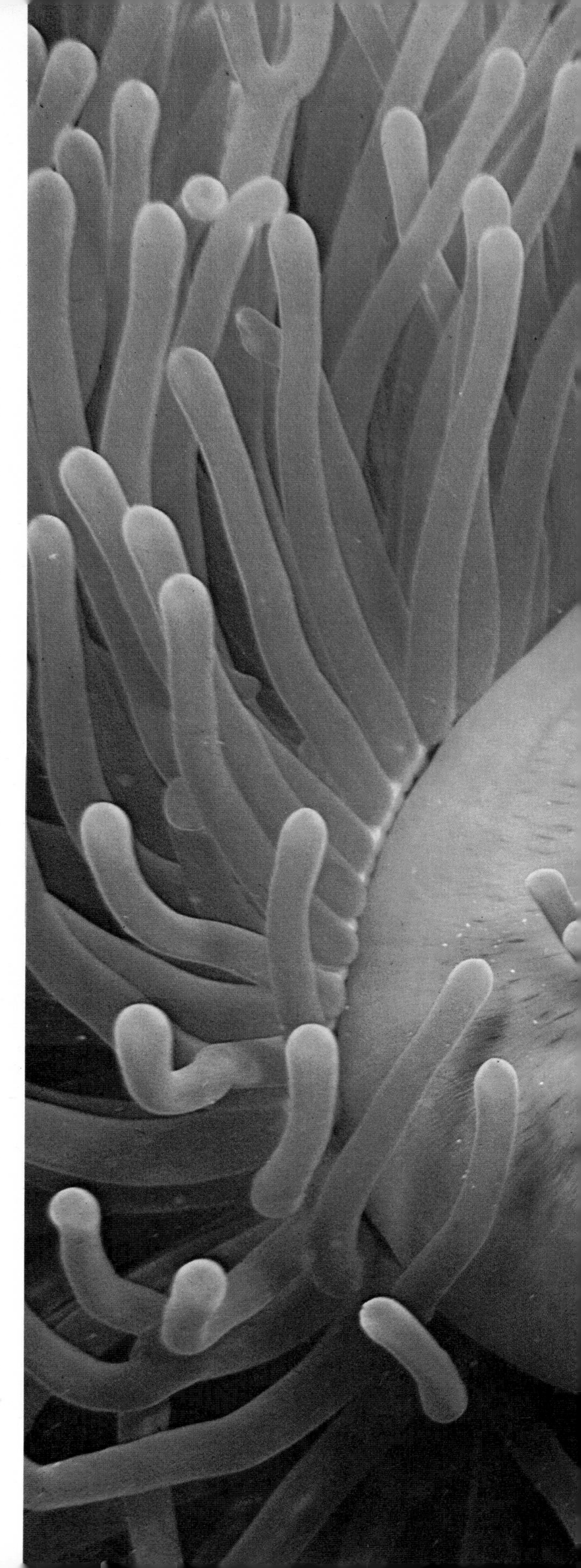

Although unable to see, smell, or hear, even a very simple marine animal needs to respond to outside stimuli. This giant sea anemone has sensitive nerve cells all over its saclike body. If touched by a potential enemy, its tentacles react by shortening and then withdrawing into the central opening.

in the water around the fish cause the fluid to move, stimulating the sensory cells and sending messages back to the brain.

In addition to the main lateral line canals, there are branches that form a network over the head and snout. Sharks and many other fish also have jelly-filled pits scattered over the head region; these probably have the same function as the lateral line.

The lateral line system is particularly well developed in fish with very small eyes. But it does more than merely help a fish to find its way about in the darkness; it also helps a fish to keep its position in a school. When a group of herrings are swimming together, the fish are able to remain evenly spaced out, with each aware of the changes in direction of its neighbors, because of the lateral line sense organ. School fish certainly use their eyes, and many live in the bright surface waters, but even a blinded fish can keep its position in the school and has no difficulty sensing when and in which direction to turn.

For many sea fish the sense of smell plays an extremely important part in their lives. In water, smelling is relatively easy because odorous substances are already dissolved (the state necessary for chemical messages to be received by the senses). Sharks and rays have very large nasal organs situated in their snouts and connected to the surface by nostrils on the underside. When a shark becomes aware of movement in the water, it makes zigzag tracking movements guided by its sense of smell and very quickly homes in on such objects as struggling fish or pieces of meat. It cannot do this if both nostrils are plugged, as scientific experiments have shown. The hammerhead shark appears to have a particularly keen sense of smell. This may be because of the placement of its nostrils: they are very close to the eyes, which are widely separated on the ends of long projections on each side of the head. Thus the hammerhead can locate with great accuracy the direction from which a scent comes.

Most bony fish have paired nostrils for carrying water in and out of the nasal sacs on each side of the snout. As the fish swims along, water passes into these blind nasal sacs, which are lined with sensory cells. The sensitivity of some of these organs is incredibly high: salmon, for example, can find their way back to the river of their birth entirely by the scent of substances in the water, even when their eyes are covered. It is probable, then, that most fish, like dogs and many other mammals, recognize members of their own species, as well as food and enemies, primarily through their noses.

Fish have a special sensory system, often visible as a lateral line on the sides of the body, which detects vibrations in the surrounding water. On this colorful reef fish, the lateral line is marked by a row of spotted scales that cover a small canal in which the sense organs are located.

The dog whelk, a large snail-like mollusk that feeds on mussels and barnacles on rocky shores, has no "nose" for smelling. However, it does have special chemical receptor cells at the base of a movable siphon that sticks out from under the front of the shell. To breathe, the whelk draws in water over its gills through the siphon. As it hunts, the animal swings the siphon from side to side, sampling the water for traces of food. When it senses food, the whelk turns in the direction in which the siphon is pointing. Thus it detects not only the presence, but also the direction, of its next meal.

However sensitive the nose and other chemical receptors are for recognition, most fish ordinarily depend on their eyes for the ultimate close-up task of spotting and capturing their food. In the majority of sea fish—except for some deep-sea dwellers—the eyes are well developed and broadly similar in structure to those of land vertebrates, although modified in some ways because of the different medium in which they are used. Consider the eyes of a herring, for instance: a herring has no need for eyelids or tear glands to keep the surface of the eye moist and clean; the water does this. Its eyes are on the sides of the head, giving a wide field of vision all round. Moreover, as in most fish, the herring's eyes move independently of each other, further increasing the field of vision. The front of the eye, the transparent cornea, is nearly flat, but behind it is a large spherical lens, which focuses light rays onto the sensitive retinal layer that lines the eyeball. Thus, both the position and the shape of the herring's eyes enable it to see clearly objects close in front of it—which is useful for sighting nearby food or enemies. To the sides, on the other hand, it can see farther and is thus aware of fairly distant movements.

Marine mollusks have eyes that are very similar to and almost as efficient as those of fish. The octopus, for example, can discriminate between many different shapes and patterns of prey animals. Each of its eyes has an iris that

Fishes' sense of smell is particularly keen. At spawning time, salmon (above) smell their way back to the river where they were born, and the long nose of the sand tiger shark (right) enables it to navigate and find food, doglike, by scent.

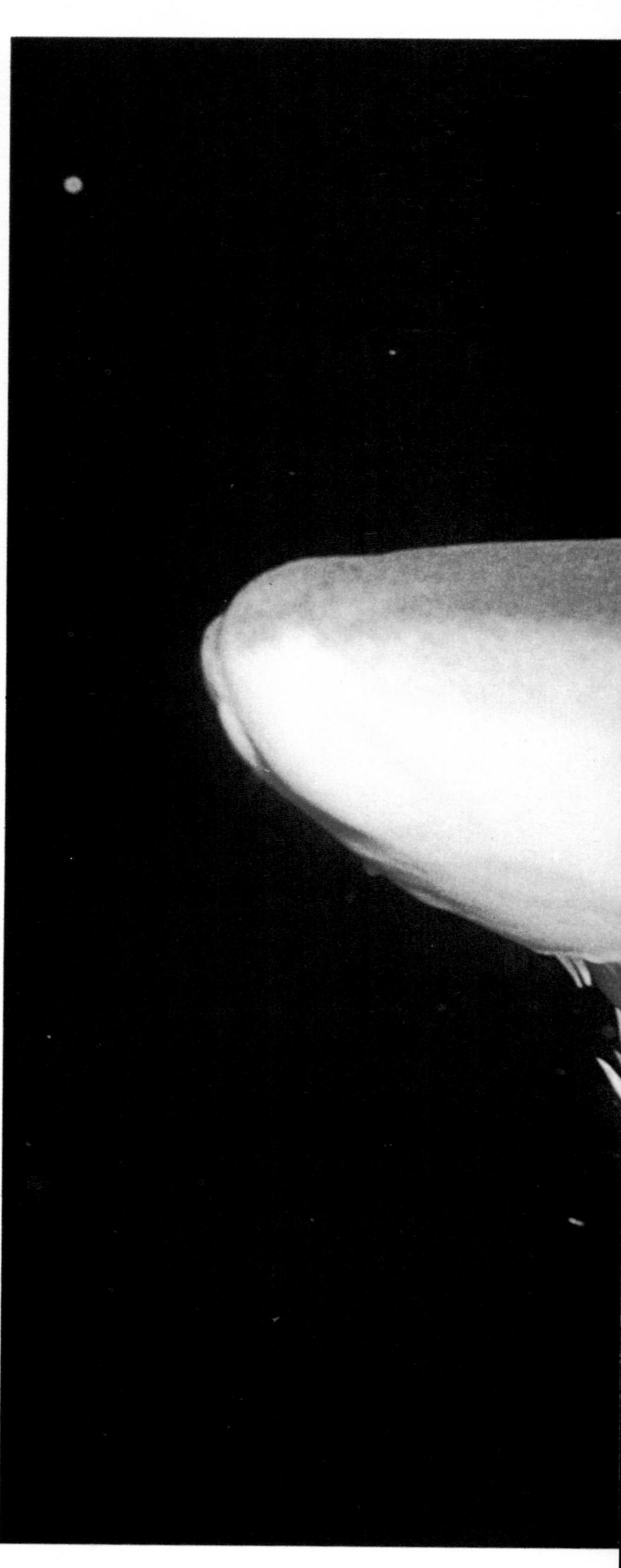

controls the light entering through the round lens and falling on the sensitive retina, and all these parts function like those of a fish's eye. The periwinkle, which browses on seaweeds, has simpler eyes but they also have a lens for focusing and a retina of light-sensitive cells. The scallop is an unusually active bivalve mollusk, and has a row of bright blue eyes all round the edge of the mantle that lines its two shells. What the individual eyes lack in efficiency they make up for in their numbers and position.

It is easy to imagine that any animal looking at the same scene as ourselves must see the same picture. This is far from true, however. Many of the lower animals see not a continuous picture but a pattern of small spots of light making up a mosaic. Marine crustaceans and worms see this type of picture through their compound eyes, each composed of many tiny separate units.

Relatively few animals see in color, but fish that live in the waters above the continental shelves and near the surface in the open oceans seem to have quite good color vision. Because only blue-green light penetrates far in the temperate and cold seas, most of the fish there have predominantly blue-green upper surfaces with

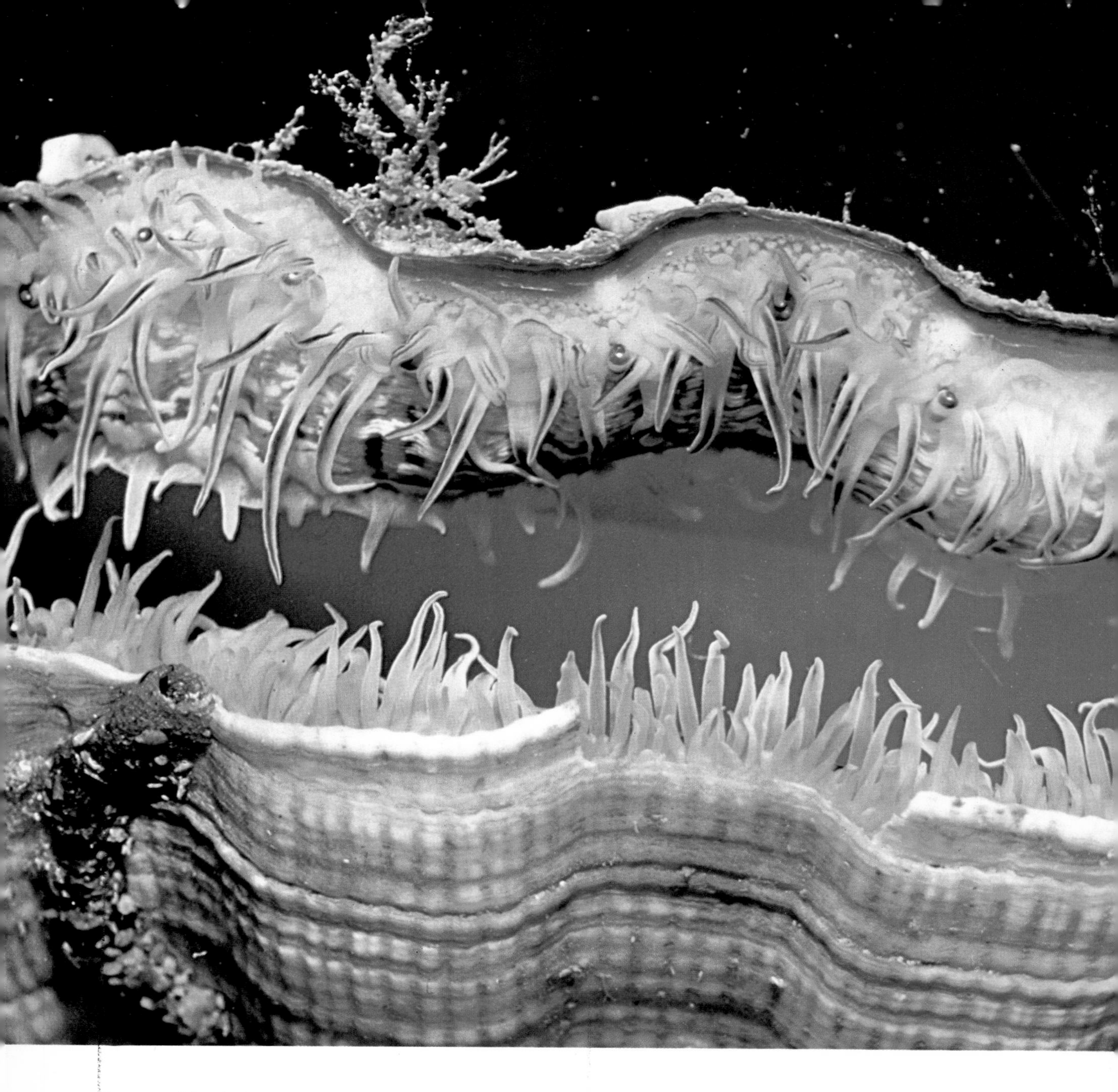

silvery white below. Those living around tropical coral reefs, however, where the clear water allows all light to pass through, display all the colors of the rainbow and presumably can see each other in these colors.

The animals with the largest eyes are those that spend most of their lives in the mid-water zone at a depth of between 300 and 3000 feet. Only a small amount of light penetrates down to this twilight region, where in clear water most objects appear green or blue. Mid-water creatures, such as the squid, have enormous pupils and lenses in relation to the overall size of the eye; this allows the small amount of light available to fall on a small area of the retina by focusing a wide beam of light into a pinpoint, thus giving the clearest possible picture. The retina in such fish is particularly sensitive to blue and green light, and there is often an extra reflecting layer, which causes the fish's eyes to shine like those of a cat caught in the headlights of a car.

Some fish living in the twilight of the mid-

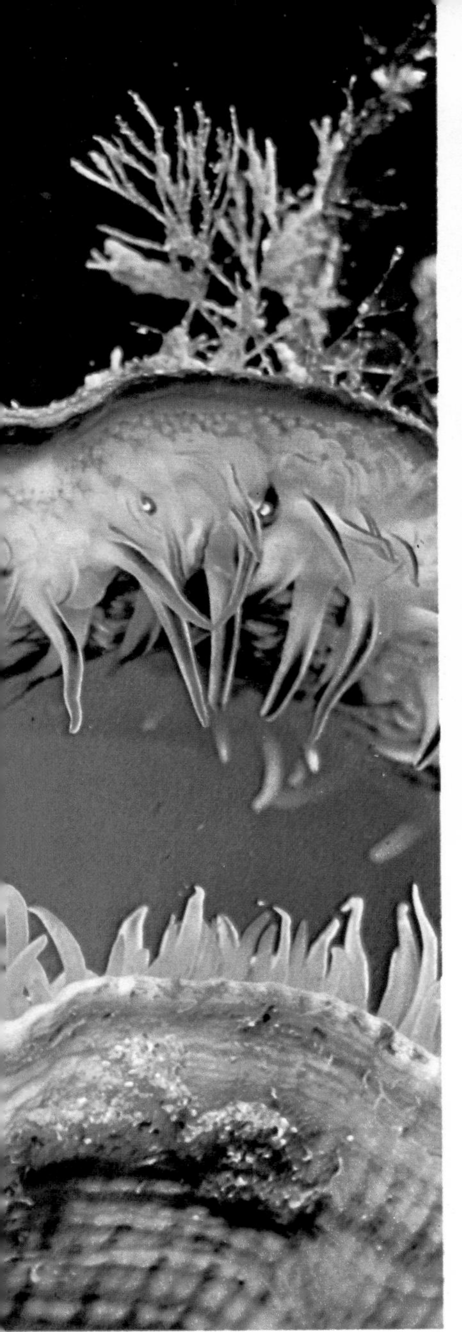

Dog whelks, seen above feeding on a dead crab, have an acute sense of smell but no nose. Instead, breathed-in water passes over chemical receptor cells at the base of the siphon, and these cells "smell" the presence and location of food.

Left: scallops, like other marine mollusks, have very efficient eyes; they appear in rows, like numerous glistening pearls, among the tentacles lining the two shells, and they function in much the same way as do those of vertebrates.

water zone have eyes that are not spherical but tubular and looking upward—from which direction most of their food and enemies are likely to come. The hatchet fish is such a creature; its tubular eyes have two separate sensitive retinas, one for seeing activity in front of the animal, the other for seeing more accurately fish that are close above it.

Finally, the animals that normally live below 3000 feet, whether swimming freely or lying on the bottom, exist in almost total darkness, and their eyes are usually small and ineffectual. The fact that they possess eyes at all suggests that they are aware of luminous objects around them, but the poorly developed eyes are only simple receptors, sensitive to changes in light intensity but deficient in detailed vision. A fish living on the seabed, at whatever depth, is more likely than a free-swimming fish to find organisms on which to feed without searching for them. Thus, the rattails, sea snails, and eelpouts have absolutely no need for big eyes, because they spend their

The big eyes of the hatchet fish enable it to take advantage of the little bit of light that reaches the middle depths where it lives, and the curved corneas give it a wide field of vision in which to spot possible food or danger approaching from all directions.

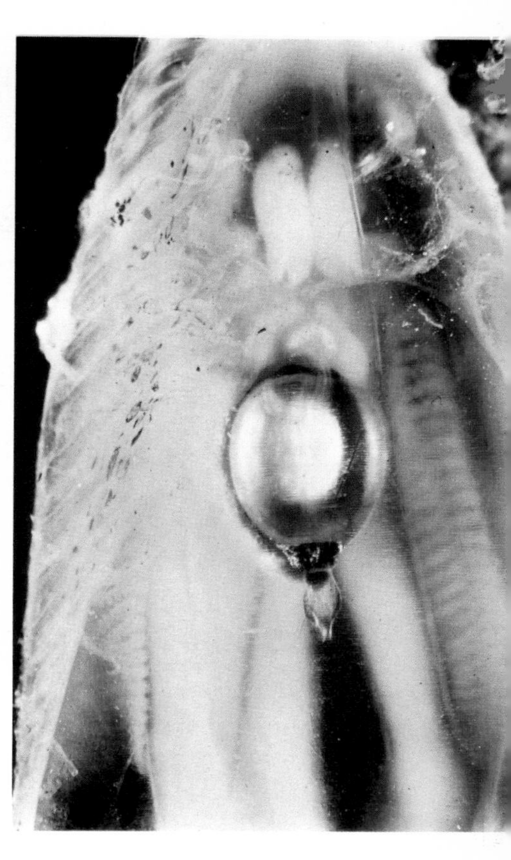

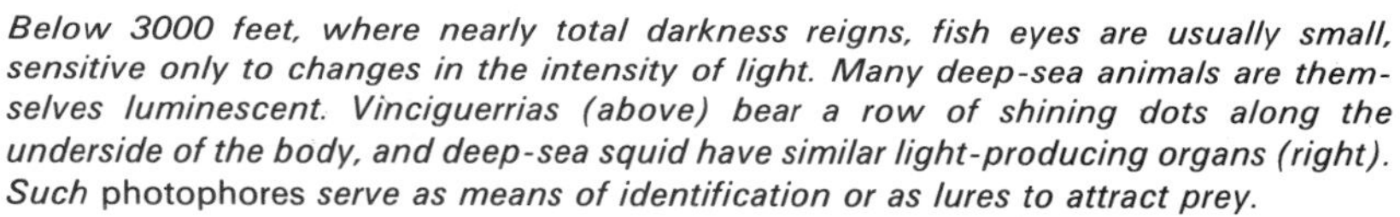

Below 3000 feet, where nearly total darkness reigns, fish eyes are usually small, sensitive only to changes in the intensity of light. Many deep-sea animals are themselves luminescent. Vinciguerrias (above) bear a row of shining dots along the underside of the body, and deep-sea squid have similar light-producing organs (right). Such photophores *serve as means of identification or as lures to attract prey.*

entire lives within one or two feet of the seabed.

Although little daylight reaches the dim regions below 500 feet, this does not mean that the animals there live in total darkness. There is much local illumination in the form of flashes or even of continuous light—illumination that emanates from the animals themselves. Divers who have explored these regions have described a variety of light-producing animals, including squids, prawns, and many fish. Most of the starfish brought up from the seabed are also luminescent. In fact, about three quarters of all the animals that swim freely at depths between 500 and 5000 feet are able to produce light in their own bodies. The most luminous depth seems to be at about 2000 feet, but flashes from living organisms have been recorded as deep as 10,000 feet.

This *bioluminescence*—to give it its scientific name—has several uses. For one thing, it illuminates the immediate surroundings of the animal so that it can see nearby food. The snout of the lantern fish, for example, bears a large luminous disk that throws a beam of light upon the krill on which it feeds. Many deep-sea fish use the luminous lures on their snouts and chins to attract smaller fish within reach of their great mouths. About 98 per cent of all the known kinds of angler fish capture their prey by means of an appendage—really part of the dorsal fin—that resembles a long rod with a luminous tip.

The light-producing fish have special organs called *photophores*. The number and placement of these luminous spots enable fish quickly to recognize members of their own species. Each kind of lantern fish and hatchet fish, for instance, has its own particular pattern of lights along the underside of the body, and these are visible from a distance of one or two feet in the dark water. Similar lights help a school of fish to keep together or solitary fish to avoid one another's territory. Not only can friend and foe be distinguished in this way, but males and females also have distinctive light signals. Male hatchet fish, although smaller than the females, have larger light organs with which they attract their mates and perhaps also guide them up into layers of water where spawning takes place.

Deep-water starfish appear to have a variety of light-producing organs. In some kinds the bases of the arms shine; in others a regular series of flashes passes down each arm toward its tip; and in some the light emanates from the organs in the central body of the animal. No definite function has yet been found for the luminosity of such slow-moving scavenging animals.

There are two ways in which living organisms can produce light. Most luminous fish have a substance called *luciferin* in special skin glands. When this is oxidized, the light released is reflected by a silvery backing layer through a small lens, to give a fine beam. The second way involves luminous bacteria, which in the case of angler fish glow in their lures. To emit light, these bacteria need oxygen, and angler fish are thought to control the amount of light by altering the blood supply carrying oxygen to the tips of the lures. Some fish have black shutters around their photophores and can either flash the light on and off like a lighthouse, or alternatively send out a steady beam. By using different patterns of light flashes and beams, the luminous fish have evolved signals designed to alarm, attract, repel, guide, or simply shed light on the other creatures that dwell in the depths.

The deep-sea fish themselves tend to be dark in color; many appear red when seen in normal light, because red is turned black by the dim blue light of moderate depths. The velvety scales of these fish reflect no light and so they are practically invisible in their natural habitat except for the light from their photophores.

Fish living on the bottom at great depths have limited vision, but they have compensatory alternative senses. Like the majority of fish, they have scales, but these tend to be thin and the skin contains many sense organs that give the body an all-over sensitivity to contact. The tripod fish "walks" about on the bottom on three long projections from the pelvic fins and tail. These contain sense organs that are highly useful both for probing and for finding food; and this sense of touch makes up for the fish's very small eyes.

Although they spend much of their lives out of water or in the well-lit upper layers, marine mammals also rely greatly on their sense of touch. When a sea lion dives, it must close its nostrils, and its eyes and ears are poorly adapted to function under water. But, like seals and sea

A keen sense of touch compensates for the limited vision of such bottom-dwelling fish as the gurnard, which creeps over the seabed on the stiffened rays of its pectoral fins; these serve as sensitive feelers probing for food, such as crustaceans.

otters, sea lions have particularly long *vibrissae* (whiskers) on the snout. These extra-thick stiff hairs are bent by slight contact and by quite small movements of the water. Messages about the altered position of the vibrissae pass back to the brain from special nerve cells in the tiny pockets in which the hairs grow. And this is how a sea lion "feels" the nearby presence of a fish and is able to chase and catch it. Most people think of dolphins and other whales as hairless mammals whose fur has been totally replaced by an insulating layer of blubber. On the snout and chin of most of them, however, there are scattered short, stiff hairs similar to vibrissae; and these, too, probably have a tactile function.

A sense of balance is very important to all animals that are active: they must be able to tell which way up they are. This sense is particularly important to animals moving in the three-dimensional worlds of water and air. Fish and whales have a sensitive organ of balance in the inner ear. In this organ is a liquid containing tiny suspended chalk particles that, under the influence of gravity, fall onto sense cells arranged in a three-dimensional system of curved tubes. The octopus has a very similar system, situated in the head. And the lobster uses sand grains, taken from its habitat, in two hair-lined pockets at the bases of its antennae.

In vertebrates, the sensitive inner ear is responsible for both a sense of balance and hearing. Although a fish has no external ear and no eardrum it still hears well. It does not need such organs, because the small pressure changes set up in water by sound vibrations can pass unhindered from the water through the equally dense body tissues to the inner ear. Experiments have shown that fish are able to detect sound over a wide range of frequencies. On the whole, small fish seem more sensitive to high-pitched sounds and large fish to lower ones.

Because fish are able to hear well, it is reasonable to assume that sound plays a significant part in their lives. If a hydrophone (an underwater microphone) is placed among a school of anchovies, it will pick up a series of rhythmic sounds while the fish are swimming actively, but not while they are still. To the individual fish, these sounds, probably produced by the sudden movement of fins against the water, may serve as an additional means of keeping in close touch with its school. They may also help to indicate the position of the school to predatory fish.

When a parrot fish is chewing on a piece of coral, its heavy jaws make loud crunching noises. These may well lead other coral-eaters to the source of the appetizing noise and so a group may gather. Other noisy eaters are pufferfish and triggerfish, which munch sea urchins, and rays, which feed on crabs and mollusks. Even some fish that eat softer food, such as the many kinds of wrasse, make rasping or clicking sounds as they eat. In all cases the sounds are probably not just incidental but appear to be used as a means of communication between members of a species. In some such way males and females of solitary species are perhaps able to find each other more easily in the breeding season.

There are many different types of sound made by fish. One example is that made by a school of herrings, feeding on planktonic animals in shallow coastal waters. When the fish swim up to the surface each one releases air from its swim bladder. The escaping air bubbles produce a high-pitched squeak, and this source of sound may also be used in communication: perhaps it enables the school to move on together.

A number of fish commonly found off the coasts of the United States, including the toadfish, drum fish, and squirrel fish, use their swim bladders to produce an impressive volume of noise. People who have heard it have variously described it as drumming, groaning, honking, growling, grunting, and croaking. Just to single out one example, the toadfish has a heart-shaped swim bladder surrounded by muscles that vibrate the bladder to make a resonant hollow sound. The noise is loud enough to be heard out of the water, and the underwater effect can be deafening. Its purpose may well be related to territorial behavior, because the toadfish seems to sound off chiefly when it senses the approach of intruders.

The dim world of the ocean deeps, where the great whales live, has been compared to the aerial world of the bats. These two very different groups of mammals share a common method of navigating by echolocation (very much like a man-made sonar system) and both groups also use sounds to communicate among themselves. The sounds used in echolocation are clicking and trilling noises, produced by altering the air pressure within the respiratory passages.

There has been much research on whale sonar systems, most of it done with captive dolphins—animals that can unfailingly detect and avoid fine-mesh nets and thin wires stretched across a

Like many other marine mammals, the harbor seal has weak eyes and no sense of smell under water—where it must keep its nostrils closed—but long, stiff vibrissae *(whiskers) help it to find its way about and locate food. Each vibrissa grows from a small pit in the skin, where sensory cells are stimulated when the hair is bent by slight contact and even by small movements of the water.*

pool. Their echolocation system consists of a series of high-frequency pulses that quicken as the dolphin approaches a solid object, from which the pulses are reflected. The dolphin scans the object by weaving its head slightly from side to side to make distance- and direction-finding more accurate. Dolphins and other toothed whales have a prominent oil-filled forehead, which can change shape and probably focuses the ultrasonic sounds into a narrow forward beam.

Such echolocation sounds are inaudible to the human ear, but the sounds of lower frequency by which whales apparently identify and communicate with one another can be heard quite clearly out of water. These sounds are probably produced in the larynx, although whales have no vocal cords. If a hydrophone is placed among a school of dolphins, it picks up a continual chatter of many distant voices. Although there is little visible on the outside, dolphins and other

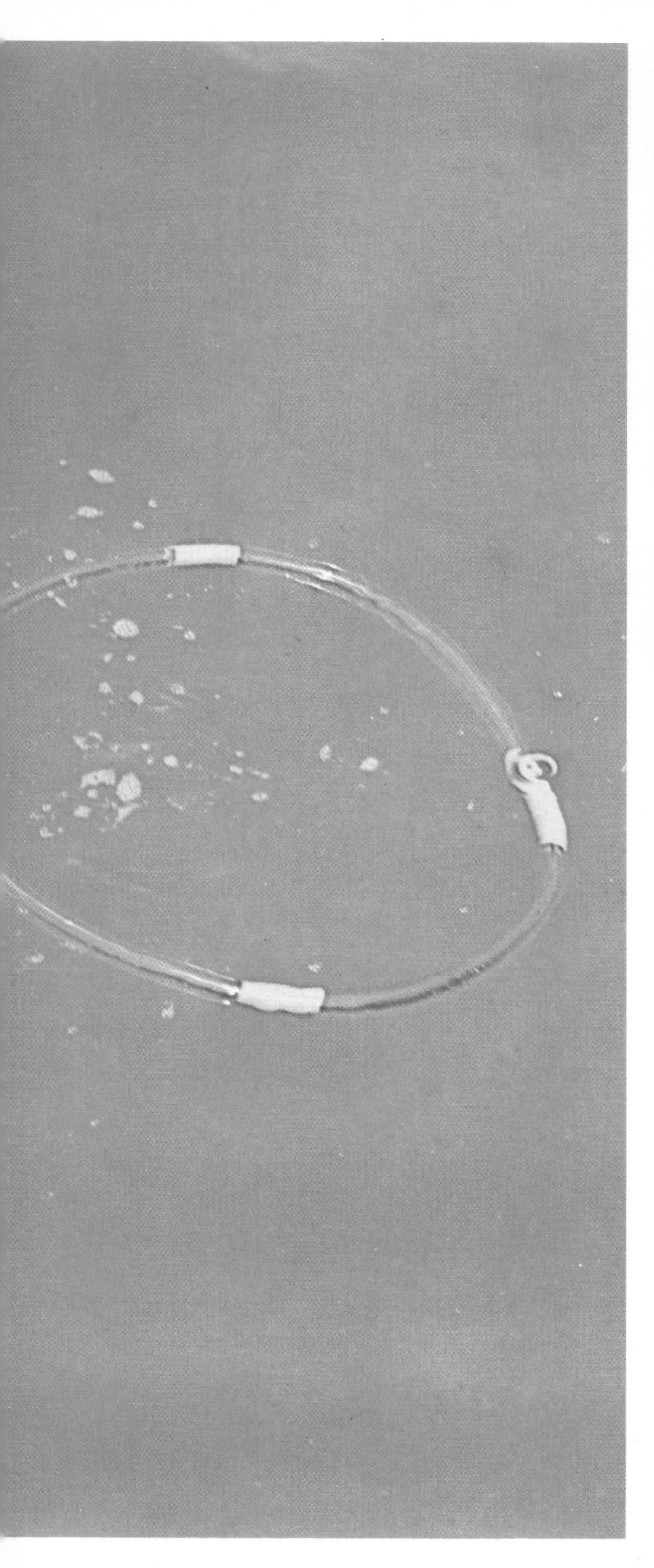

Above: fish are neither silent nor deaf. When the triggerfish crunches coral, it makes loud noises, and such sounds are probably a form of communication—either to direct others to the food source or to signal the fish's territory.

This dolphin's very accurate sonar system allows it to cope with such floating obstacles as wire rings even when it is blindfolded. Dolphins and other whales do their navigating and communicating by making high-frequency squeaks that bounce back off solid objects; sense organs in the lower jaw pick up the echoes and transmit them to the internal ears.

whales have very sensitive ears, but they receive sounds first through special organs in the lower jaw. The vibrations are then transmitted to the middle and inner ear through the jaw bones.

Thus, we have seen that the familiar senses of smell, sight, touch, and hearing play an important part in the lives of saltwater animals. Other senses, of which we can have no experience, are also well developed in many of them. As with land animals, the keenest senses are found in the most active animals, which need to find their way about at speed. Slow-moving and sedentary animals can make do with a general sensitivity all over the body surface.

Moving in the Sea

The majority of animals must move about to find food; only a few can lead sedentary lives, waiting for food to come to them. Movement in the sea is aided by the fact that water is about 800 times as dense as air, and can therefore support much greater weights—up to 120 tons in the case of the blue whale. The density of water also means that although no effort is needed to support the body's weight, considerable effort is needed to start it moving. Once an animal has begun to move in water, however, little more muscular work is required to keep it moving—freewheeling is easy.

Streamlining is an important feature of many marine animals. It illustrates the biological principle of convergence, in which unrelated or distantly related animals look alike because their ways of life are similar. A shark, a tuna fish, a penguin, a seal, and a dolphin all have very much the same basic shape, because they all live and feed in the sea.

A smooth outline with no protruding edges is typical of good streamlining, but it is not the whole story. When an animal moves through water head foremost, the friction between the water and the animal's shape causes eddies, leading to drag. A perfectly streamlined shape, rounded at the head, thickest about one third of the way back, and tapering smoothly to a narrow tail, gives the maximum reduction in turbulence at the body's surface, and therefore the least wastage of energy.

We find this tapering or *fusiform* shape in many fish but by no means all. There are fish with bodies flattened from top to bottom, such as the rays and skates, or from side to side, such as plaice, sunfish, angelfish, and many more. Flatfish are mainly sluggish bottom-dwellers that feed on the slow-moving or sedentary animals of the sea floor, and they have sacrificed speed for effective camouflage. A flattened shape does not necessarily make a fish slow or clumsy, as you can see if you watch the quick, darting movements of a little butterfly fish, or the graceful way

Streamlining conserves energy. The unbroken gliding movement of dolphins as they surface to breathe is made possible by the momentum produced by their streamlined bodies.

in which a skate glides along. Fish whose shape lacks all streamlining, such as puffers (or globe-fish) and sea horses, usually have particularly effective armor-plating or camouflage to compensate for their lack of speed.

In marine animals, as in ships, the best position for the main propulsive force is at the back. Thus, most fish use their tails to drive them through the water. As the muscle blocks attached to the flexible backbone alternately contract and relax, pulling the tail first to one side and then to the other, water is pushed backward and to the side and the fish moves forward. The tendency of such a movement to cause the head to swing from side to side is balanced partly by the weight of the head and partly by the fins at the front of the body. In fact, a series of waves passes from head to tail down the body of any fish as it swims, but we can see these waves clearly only in long, thin fish such as eels. In a mackerel they are less visible, and in a tuna fish, which has a relatively deep body, we can hardly detect them.

In addition to the tail fin, fish have single fins along their backs and on the underside, as well as paired fins, and any of these may be used in movement. Wrasse row themselves along with their front paired fins *(pectorals)*, and normally use the tail only as a rudder. The little sea horse uses its small dorsal fin to propel itself from one piece of weed to the next, whereas the sunfish, among others, uses synchronized movements of both dorsal and anal fins. The enormous pectoral fins of skates and rays allow them to "fly" through the water, unaided by the thin whiplike tail.

Fish, however, are not the only animals that use a finlike structure as a propulsive force. All

Below: the pufferfish cannot drive itself through the water by its tail, as other fish do, because its body is rigid. Pufferfish have sacrificed streamlined speed for the protection afforded by their tough, boxlike armor of bony plates.

Although most fish are alike in having tapered bodies and in using their tails for propulsion, they have evolved a number of different methods of steering, balancing, and depth control. The bony mackerel (above), made lighter than water by its gas-filled swim bladder, can change directions by means of its flexible fins. The dogfish (right) has a flattened head and stiff paired fins, which help to counteract the downward movement produced by the muscular tail. Because the dogfish is denser than the water, it sinks if it stops swimming.

Like all rays and skates, this bat ray spends much of its time lying on the seabed, but it can also "fly" through the water by flapping its winglike front fins in an undulating motion.

whales have horizontal flattened tail flukes, which correspond to the tail fin of a fish. When a whale swims it bends vertically instead of horizontally. Seals swim by bending the hind part of the body from side to side, using their short hind limbs and webbed feet, which point permanently backward, to provide the driving force. Sea lions, which spend more time on land than seals, swim by means of their short flattened forelimbs, using their hind limbs as rudders in the water; they can turn the hind limbs forward to help them waddle along on land.

Although all the fastest and best-streamlined aquatic animals use their tails for swimming, the movements of the tail fin are another cause of turbulence and drag. To reduce this as much as possible, the tail fin is often forked, as in the herring and mackerel, or sickle-shaped; the latter is the typical tail shape of the mackerel's close relatives, the sailfish and swordfish. These are very swift swimmers. In fact, the greatest speed ever claimed for a swimming animal was 68 miles an hour by a sailfish. And its relatives the wahoo (or peto), marlin, swordfish, and bluefin tuna are said to reach speeds of between 40 and 50 miles an hour on occasions.

Besides having almost perfect streamlining, members of the mackerel family have thin, keel-like dorsal fins, which can be folded back when swimming at speed, and relatively small paired fins. They also have extraordinarily powerful muscle blocks in the back and tail region. Scales may be only a minor obstacle to speed but they do cause a small amount of drag, and it is interesting to note that many fish of the mackerel family are almost scaleless. It is also a fact that tuna are the only fish known to be able to regulate their body temperature. This means they can keep their body functioning at the high rate necessary for sustained bursts of fast swimming.

Compared with these very fast swimmers, the speeds attained by other marine animals do not seem very remarkable. Salmon can swim at 15 miles an hour, trout at 7 miles an hour, and herrings at 4 miles an hour. The fastest penguins are the Gentoo and Adélies, which can reach speeds of about 20 miles an hour when fleeing from a predatory leopard seal. Among the marine mammals, the fastest are the dolphins at 35 miles an hour.

Besides speed under water, which is usually associated with chasing food or escaping from a predator, a number of marine animals are able to make prodigious leaps out of the water. As one might expect, the fastest swimmers have the necessary body shape and propulsive force to enable them to do this. Sailfish, tunas, dolphins, manta rays, and whales are the most spectacular leapers in the oceans—but nobody has yet clearly understood why they leap.

Although a fish can move with little effort, it needs efficient means of steering, balancing, and depth control. These are achieved in quite different ways in the cartilage fish (such as sharks) and in fish with bony skeletons (such as herrings and cod).

Sea lions are awkward waddlers on land, but they swim with agile grace in the water, where they propel themselves along by means of short, flattened forelimbs, using their hind limbs as rudders to steer a course among the rocks.

A shark's body, which is made of dense tissues throughout, is heavier than water and will sink to the bottom unless compensated for in some way. Furthermore, the tail is asymmetrical, with a large upper lobe and a smaller lower one; and although this tail shape gives a strong driving force it also results in the lifting of the tail and a corresponding lowering of the head of the fish. Counterbalancing this tendency a shark has large pectoral fins that are slanted to exert a lift like that of a bird's wings, and a flattened head end, which also helps. The paired fins are widely separated, broad, and more or less rigid, and so they are of little use in steering. The median fins are also rigid and prevent the fish from rolling sideways. To maintain its level in the water, therefore, a shark must keep moving and steer with its tail. This is why large ocean-going sharks are never still.

Bony fish are commonly seen hanging motionless at various depths in the water. Most living bony fish have a large, thin-walled internal sac called a *swim bladder*, which enables them to control their density in relation to that of water. A herring's swim bladder is like a small, silver balloon and it is connected with the fish's throat by a thin tube. By swallowing air at the surface or expelling it through its mouth, the herring alters the pressure in its swim bladder and thus regulates its position in the water. It takes no muscular effort for the herring to maintain this position, and so its thin, flexible fins can be used entirely for steering. Because of the need to recharge an open swim bladder at the surface, such bladders are found only in fish living on the continental shelf.

The bottom-feeding cod has an even more efficient system, based on a completely closed swim bladder—one that is not connected with the throat. The pressure within this type of bladder is altered by gas secreted into it or absorbed from it by the blood, and there is thus no need for the cod to swim to the surface. This type of bladder also occurs in deep-sea fish. The paired fins of the cod are very close together, with the pelvic fins (normally toward the rear of the body) slightly in front of the pectorals. This

The great blue marlin, or sailfish, is notably swift, reaching speeds of 50 miles an hour, and often makes spectacular leaps out of the water. Its superb streamlining, folding dorsal fin, and nearly scaleless body reduce drag, and the strong muscles and sickle-shaped tail fin add force as it speeds through the water.

The 12-foot-long blue shark lives in apparent conflict with physical law: although denser than water, it can maintain its level in the water and never sinks, because it never stops swimming. And despite its size, it uses up less oxygen than almost any other fish.

gives them great maneuverability, and so cod are among the most agile of fish.

Swimming is not the only method of moving in water; there are other ways involving much less energy. Lobsters and crabs crawl about on the bottom on their jointed legs. When alarmed they swim quickly backward to hide in a crevice.

An octopus normally moves about slowly on its long tentacles but it also has a powerful jet propulsion system for emergencies. By squirting water backward from its siphon it can shoot out of sight with its arms trailing behind it. Squids also use jet propulsion for swimming fast after their prey. Some large squids have been reported swimming steadily at 23 miles an hour, and smaller flying squids swim fast enough to shoot out of the water and glide above the surface.

Sea snails, such as winkles and whelks, glide over the surface of rocks and weeds on their muscular feet. Waves of muscle activity pass

forward along the bottom of the flat foot. Some of these animals can move quite rapidly as two waves, slightly out of phase, pass along opposite sides of the animal.

A different type of gliding movement can be seen if you look closely at a starfish or a sea urchin. These animals have thousands of tiny suckers called *tube feet* arranged in rows among the spines of the hard shell. The tube feet are operated by fluid pressure and form a continuous clinging belt on which the animal can move in any direction. They are also used to open the bivalve mollusks on which starfish feed. Working in relays, the tiny suckers pull the shells apart by tiring the muscles that are holding them together.

Sea anemones and corals are relatively inactive animals, and apart from moving their tentacles and contracting or extending the body, they rarely move about after their larval development. Related jellyfish, however, are quite mobile and are often seen swimming in the surface waters. Rhythmic pulsating movements of the elastic bell-shaped body are brought about by a ring of muscles around the outside of the bell.

There are quite active swimmers, too, among the microscopic organisms of the plankton, many of which have legs, cilia (hairlike processes), or tentacles to help them move. The tiny animals of the plankton are often carried passively by water currents but most of them make regular vertical migrations between the surface of the sea and depths of 100 feet or so. They rise to the surface at night and sink during the daytime—and various theories have been put forward as to why and how they do it. Perhaps they respond to changes in light intensity, oxygen content, or the temperature of the water, perhaps to the need for food or to a built-in daily rhythm within the organisms themselves, or maybe to a combination of all these factors. Whatever the reason for their daily journey up and down, planktonic animals must have enough buoyancy to float at night before slowly sinking during the morning, and they must also have enough power to swim up to the surface again in the latter part of the day.

Various factors contribute to the extreme buoyancy of plankton organisms. The minute size of most of the animals and all of the plants gives them, for one thing, a very large surface area in proportion to their weight. The body surface is further increased by spines, hairs, fringed legs, and other projecting structures. In addition, many of these organisms are shaped like stars, needles, disks, or ribbons. And in contrast to their relatives living on the seabed or in deep water, many planktonic organisms have very thin light-weight shells. Some contain structures to give them extra buoyancy, such as gas bubbles in the protoplasm, oil droplets, or air bladders like miniature water wings. All these features help to

Flying fish only seem *to fly! Actually they escape from predators by launching themselves out of the water by means of a force produced by the tail. This permits them to glide for several yards, with the widespread pectoral fins (kept folded when in water) acting as sailplanes.*

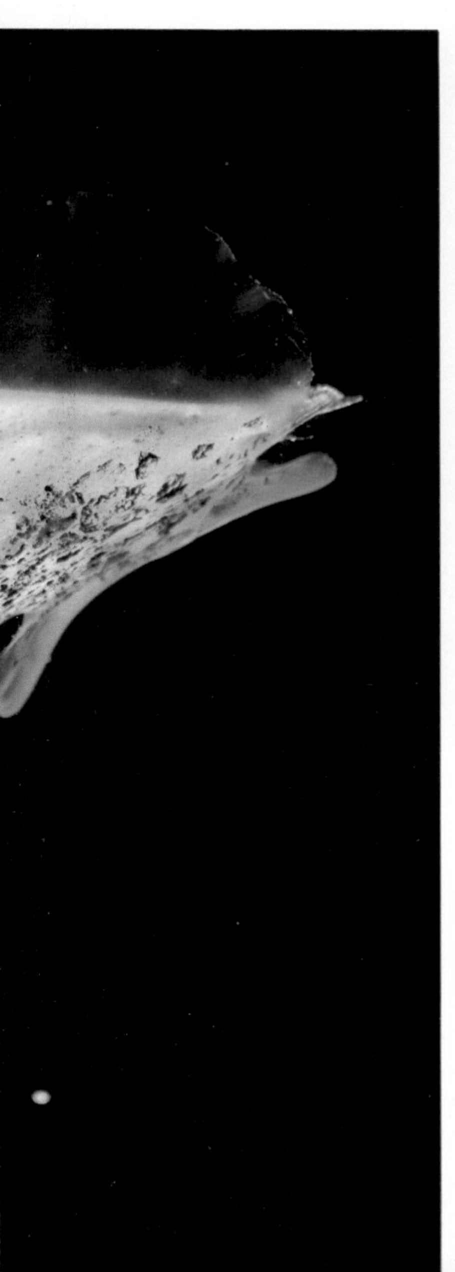

Left: the principle of jet propulsion governs the movements of the squid, which is an open-water mollusk. Water is drawn into the body and then forced out as a jet through the siphon, which is just behind the eyes. This causes the animal to shoot backward, trailing its tentacles, at speeds of 20 miles an hour; its large and efficient eyes are constantly on the lookout for prey, and for the many predators that feed on squid.

The carnivorous common whelk glides slowly over rocks and seaweed on its flat, muscular foot. Coordinated waves of contraction passing along the bottom surface allow it to move at a speed of as much as 24 inches a minute.

slow down the rate at which they sink, rather like a parachute or the hairs of thistledown. Only a few of the larger, planktonic crustaceans, fish, and jellyfish have to keep afloat by swimming.

Another group of animals for whom floating poses little difficulty is the seabirds that catch their food by diving from the water's surface. A feature of all birds is the possession of hollow, often strutted bones, which considerably lighten the skeleton. The covering of feathers further increases their buoyancy because of the air trapped among them; and birds also have air sacs that penetrate from the lungs into many tissues, including the bones. Such properties, which help to make flight possible, keep a diving bird afloat even in rough seas.

When it dives, however, a bird such as a penguin must temporarily make itself heavier than water. This is done partly by the powerful thrust from the webbed feet that all seabirds possess, and from the wings that birds such as penguins use when swimming under water. In addition penguins have more solid bones than the majority of birds. They further reduce their buoyancy by having small, closely packed feathers. And to compensate for the lack of insulation they have a thick layer of blubber beneath the skin.

A close-up of the underside of a starfish shows the tiny suckers, or tube feet, arranged in rows upon its shell. Operated by fluid pressure from within, they form an adhesive conveyor belt on which the starfish glides smoothly in any direction. It can also use its tube feet to force open the shells of the mollusks on which it feeds. It then extrudes its stomach and digests its prey.

Birds' feathers give buoyancy and insulation in the sea only as long as they are waterproof, for waterlogged feathers are as heavy and cold as wet fur or clothes. All birds spend much time preening to straighten their feathers and to coat them with oil from the preen glands at the base of the tail.

Seabirds are not the only air-breathing animals that dive to feed. Of the three groups of diving marine mammals—seals, whales, and sea cows—only seals emerge onto the land, and in general they make the shortest and shallowest dives. A seal usually remains submerged for up to five minutes and reaches a depth of between 100 and 250 feet. Dives lasting up to an hour have been recorded for the Weddell seal, which is able to go as deep as 1000 feet; but this species is unusual among seals because it feeds near the bottom on squid, fish, and crustaceans of the Antarctic continental shelf.

Seals, like all mammals, breathe air by means of lungs and must therefore have special ways of dealing with the two main problems of diving mammals: an adequate supply of oxygen for the

tissues, and the effects of water pressure. One of the functional adaptations of these animals is the slowing of the heartbeat that occurs immediately on diving. From a normal pulse of between 50 and 150 beats a minute, the rate drops to between 4 and 10 a minute, and returns to normal only when the seal comes up to breathe. The reduced heartbeat means that the blood circulates more slowly and so the oxygen in it lasts longer. In addition, because its body contains a relatively large volume of blood and has especially adapted muscles, the seal can carry sufficient oxygen to keep it going during a dive.

The rate at which the seal's body functions during the dive is also very slow. Its tissues can build up an "oxygen debt," which is paid off when it next comes to the surface to breathe. Until this happens the muscles are able to work without oxygen. In addition, the part of the brain that controls breathing is less sensitive to the build-up of waste gases in the body and so the seal does not have to breathe as often as a human diver does. Seals spend much time lying basking on rocks; while doing this they are probably paying off a large oxygen debt, as well as allowing the digestive system to deal with the food swallowed on the last dive.

When a human diver returns to the surface from a depth of only 40 to 50 feet, he will get "the bends" (caisson disease) unless he ascends very slowly. This is because nitrogen in air breathed under pressure is forced from the lungs to dissolve in the blood. When the outside pressure is reduced, the nitrogen comes out of solution in the form of tiny bubbles, and these can block small blood vessels, causing permanent disability or even death. The main reason why seals do not suffer from the bends is that they do not breathe air under pressure. When a seal dives it carries down with it only a small volume of air in its lungs, because it breathes out just before diving in order to reduce its buoyancy. The pressure of the water makes the lungs collapse, and the air is pushed into thick-walled air passages where nitrogen cannot pass into the blood. Thus, seals can make long dives to feed or to escape from enemies such as polar bears and men, and can return quickly to the surface with no danger of the bends.

Whales are the most completely adapted diving mammals; they never leave the water, as seals must from time to time, and in many cases they probably never even come within sight of land. The longest recorded dive of a whale is by a sperm whale, the largest toothed whale, whose dives, down to a depth of nearly 4000 feet, can last up to one hour and 50 minutes. Porpoises, dolphins, and whalebone whales are adept divers

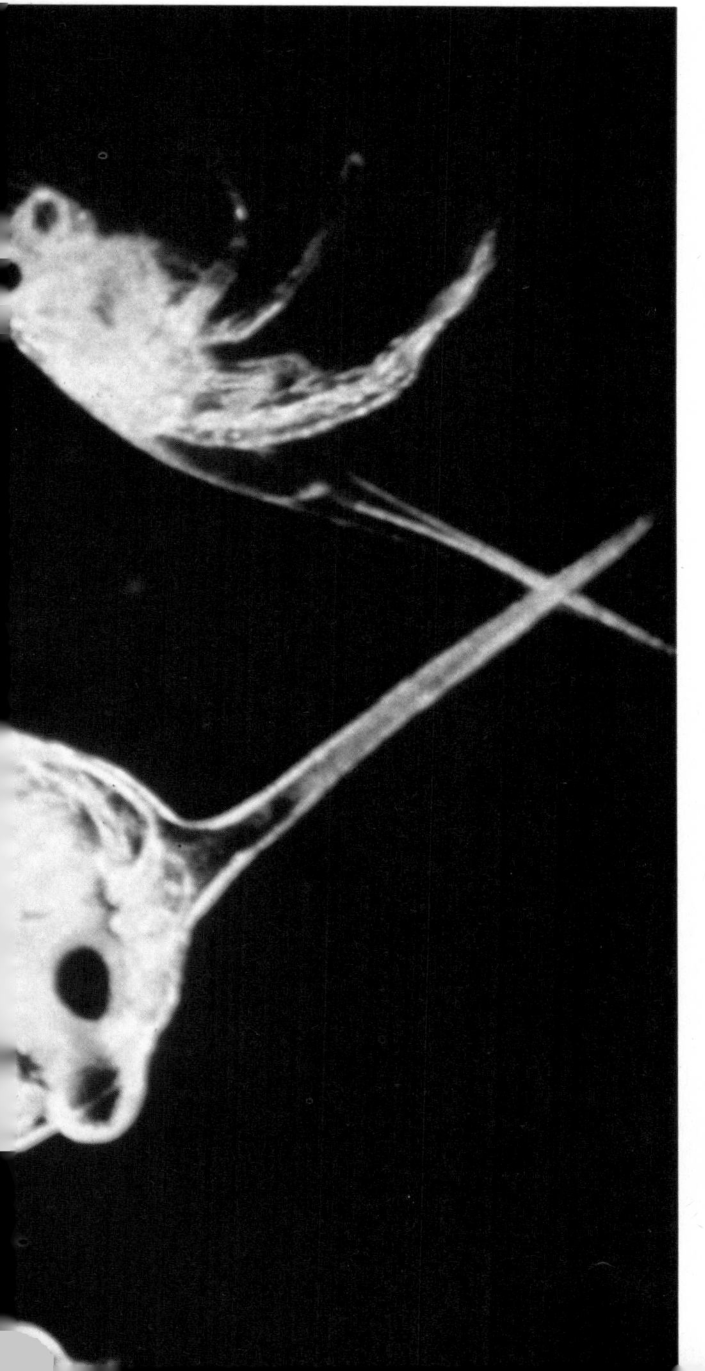

Although most of the microscopic animals of the plankton spend much of their lives drifting near the surface, many of them move downward during the daylight hours and upward at night. The buoyancy that permits them to float is partly due to their large surface-to-volume ratio, which is further increased by spines, hairs, and other projecting structures.

too: they usually stay under for about 15 minutes and go down to between 1000 and 1500 feet.

Whales normally have a slow heartbeat rate of between eight and 24 beats a minute, depending on the species, and this is apparently not reduced on diving. Like seals, they also have a very large carrying capacity for oxygen in the blood and muscles and can tolerate a high concentration of waste gases in the blood before they are forced to breathe. As in seals, too, their muscles can work without oxygen during a dive and so build up an oxygen debt. In the large toothed whales a high oxygen debt is paid off by numerous spoutings while at the surface; a sperm whale, for instance, blows out 30 times in about 10 minutes. The blue whale may blow only five times in five minutes. When the whale spouts it is simply breathing out warm air that condenses when it meets cold air.

Although the whale's muscles can do without oxygen during a dive, this is not true of the nerve tissues in its brain and spinal cord. Around these vital organs are thick networks of capillaries serving to increase the blood pressure—and thus the supply of oxygen to the nerve cells—while blood pressure in other, less vital parts of the body remains low.

Despite the great depths to which whales generally dive, there is as little danger of their suffering from the bends as there is for the seals, and for much the same reasons. From the start of the dive there is relatively little air under pressure

Penguins, the best adapted of all birds for living and feeding at sea, are equally at home under water (left) or afloat. Their small, lightly packed feathers trap air for buoyancy and warmth, and they use their wings for swimming under water. Because their bones are more solid than those of most birds, it is easier for them to remain submerged under the surface of the water.

within the whale's lungs. The proportion of lung capacity to body weight in a whale is probably the lowest for any mammal; much of the air taken down passes into thick-walled breathing passages, and so retains all its nitrogen. Moreover, whales have two further adaptations for dealing with water pressure at great depths. First, the diaphragm is oblique, not at right angles to the backbone, as it is in most mammals. Because it is attached to the ribs higher at the front than at the back, this helps to collapse the lungs during a dive. Secondly, the air sinuses in the whale's head are filled with an oily froth in which nitrogen quickly dissolves instead of passing down into the lungs. When the animal spouts, therefore, it can rapidly expel this nitrogen before taking in fresh air.

Diving mammals not only have special arrangements of both structure and functioning for holding their breath for long periods under water, but they are also equipped to make the best possible use of the time spent at the surface. Whales and seals have unusually efficient ways of pumping air in and out of the lungs. Partly because the lungs are so small, they can be filled and emptied almost to capacity by deep breathing. In comparison, man cannot normally change even 50 per cent of the air in his lungs in one breath. There is no shallow breathing, as there is in terrestrial mammals, when the animal is resting or inactive. As a result, whales and seals are able to utilize a high proportion of the oxygen from the air they breathe.

As we have seen, a seal blows air out of its lungs quite violently to reduce its buoyancy just before diving. Then it closes its nostrils, which are on the very end of the nose, by relaxing the small muscles around them. Under water the nostrils stay closed with no further effort, helped by the external pressure of the water. When the seal resurfaces, it actively contracts the muscles to open the nostrils. The end of the nose always breaks the water surface first.

The nostrils are found in a similar position in all aquatic air-breathing mammals. In whales they lie on the top of the head—which is, again, the part that surfaces first. Toothed whales have a single blowhole, whalebone whales a pair. Whalers have only to see the appearance and angle of the spout when a whale blows to identify the animal accurately. For example, a blue whale blows straight up and a sperm whale obliquely forward. On each side of the breathing passage of all whales are complex valves that help to keep the blowhole closed under water.

Breathing is essential to all animals, whether their supply of oxygen comes from water, which contains about 5 per cent oxygen, or from air, of which about 20 per cent is oxygen. This gas is used in the tissues to release the energy contained in food—energy needed for muscle and nerve activity, growth and repair of tissues, and other vital functions. Whales, seals, and penguins

breathe in air and their diets have a high fat content. This rich source of energy, together with their efficient breathing mechanism and good insulation, allows these warm-blooded animals to stay warm and to be active.

All the animals that breathe in water are cold-blooded, and the limited supply of oxygen available to them is enough for their relatively inactive lives. Fish generally spend much time free-wheeling rather than swimming very fast, and most marine invertebrates seldom move fast at all. Many animals are in fact carried about by the sea itself and expend little energy on propelling themselves.

Above: a killer whale breathing out warm air, which condenses as it meets the cold. When whales surface to breathe, the nostrils surface first, for they are on top of the head. After a dive, the whale must spout several times before it can go under again.

In the freezing winter of Antarctica, the Weddell seal survives only by staying in the relatively warm water for weeks on end. In order to breathe, it opens a series of air holes in the ice, to which it regularly returns for a breath of fresh air (left). As the holes freeze over, the seal uses its strong teeth to scrape away the ice from underneath (above). Old seals whose teeth have become worn are consequently in danger of drowning.

Polar and Tropical Waters

Living in the extreme conditions of the polar regions is hard enough for warm-blooded animals; it is far more so for cold-blooded ones. The air temperature in these regions varies from an average maximum of 50°F in summer to a minimum of about −112°F in winter. The temperature of the sea water, however, is less variable, remaining just above its freezing point of 28°F all the year round. Winter conditions on land or ice, characterized by very low temperatures, strong winds, and extreme dryness, make it impossible for any cold-blooded animal, except for one or two kinds of insects, to survive. In the sea, life is rather less hard and this is why all the vertebrates found in the polar regions spend much or all of their lives in water. Most of the invertebrates live on the seabed where conditions are more or less constant all the year round.

During the long winter months in the Arctic and Antarctic, daylight is scarce or lacking and much of the light and heat from the sun's rays is thrown back by reflection from the snow and ice. Therefore, plant life cannot function and the surface-living phytoplankton remains dormant, unable to grow or reproduce. Pelagic fish and other small plankton-eaters are also forced into a state of suspended animation, or must survive on their stored food reserves and the limited amount of plankton available. The only animals that can keep going normally under these difficult conditions are carnivorous mammals and birds that can find their food in the sea and that also have an emergency food reserve of thick fat.

In the short weeks of the polar summer, the land areas heat up—quite quickly in the Arctic, more slowly in the Antarctic, with its permanent central ice cap. The surface layers of the polar oceans also receive light and heat, which melts the ice, and plentiful supplies of nutrient minerals are raised to the surface by the upwelling warm currents that replace the heavier cold water. The planktonic animals and plants enter

In the Ross Sea— part of the southernmost ocean surrounding Antarctica—the cone of Beaufort Island rises above the floating pack ice. Most plants and animals can flourish in such polar regions only during the brief summer season, when upwelling warm currents carry nutrients to surface waters.

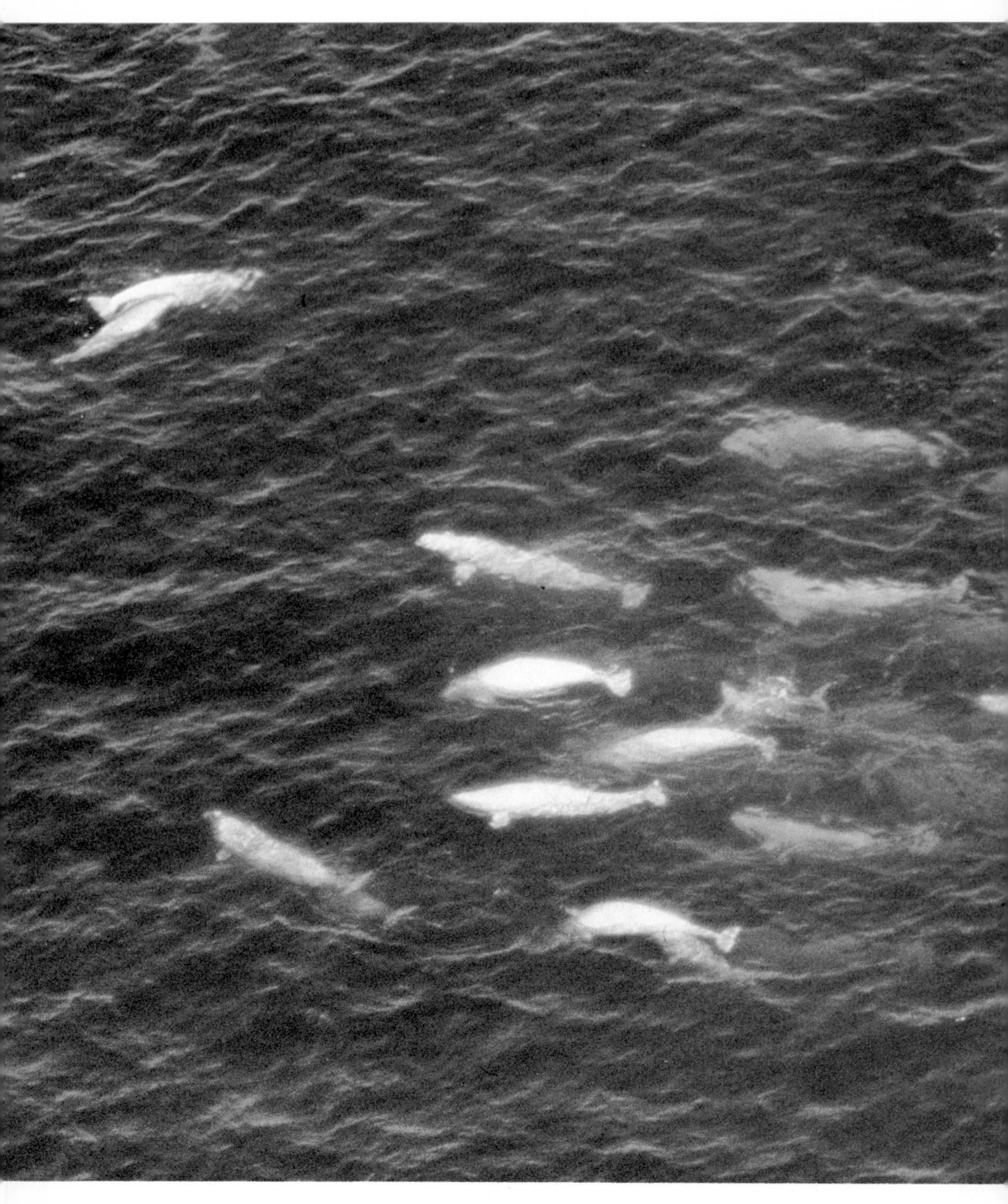

a short period of great activity, growing and multiplying rapidly, so that there is suddenly a rich food supply for all plankton-eaters. Fish, mollusks, and crustaceans grow and reproduce actively, taking advantage of the planktonic boom. At the top of the food pyramid, the carnivorous mammals, feeding almost continuously, lay down thick layers of fat under the skin to prepare them for the coming winter, which is never far away.

Polar organisms, from mammals down to simple sponges, are typically found in very large numbers, but few different types are represented. One reason for the limited number of types of cold-blooded animals in the polar oceans is that these creatures mature slowly. During one polar summer season of activity, many planktonic organisms pass through only a single generation and they survive through the winter that follows only as resistant eggs. The fewer the generations, the slower the rate of variation leading to evolutionary change, because recombinations and mutations of characteristics take place during the sexual process.

Animals living in cold environments tend to be large. This is true not only of the mammals and birds, but also of the invertebrates, many of which lay eggs with large yolks to speed up the growth of the developing young. During the short period each year when growth is possible, these animals develop from embryo directly to adult form without an intermediate free-swimming larval stage. Without these larvae, they have little opportunity for dispersal in the plankton, because the adults are more or less sedentary bottom-dwellers. As a result, in both the Arctic and the Antarctic it is common to find enormous colonies of invertebrate animals such as mollusks, worms, and starfish. Sponges, feather starfish, and scallops can be dredged up by the ton from small areas of the seabed, and acres of rocks may be covered by mollusks.

During the summer months in the Antarctic, enormous numbers of krill occur in the seas as a result of the rich supply of phytoplankton on which they feed. These concentrations of krill attract large numbers of humpback whales,

White whales gather in Hudson Bay in summertime, when the squid and fish they prey on are plentiful because of the increased numbers of crustaceans, fish, and mollusks that squid and fish eat. These, in turn, feed on the tiny, shrimplike krill, which abound in the plankton. Plant plankton complete the chain.

Although they often leave the water to sunbathe, seals must feed in the sea. The Antarctic Weddell seal (above) dives for mollusks in coastal waters; despite their name, crabeaters (right) do not eat crabs. They feed on planktonic krill, which they strain from the water through their triple-pointed teeth.

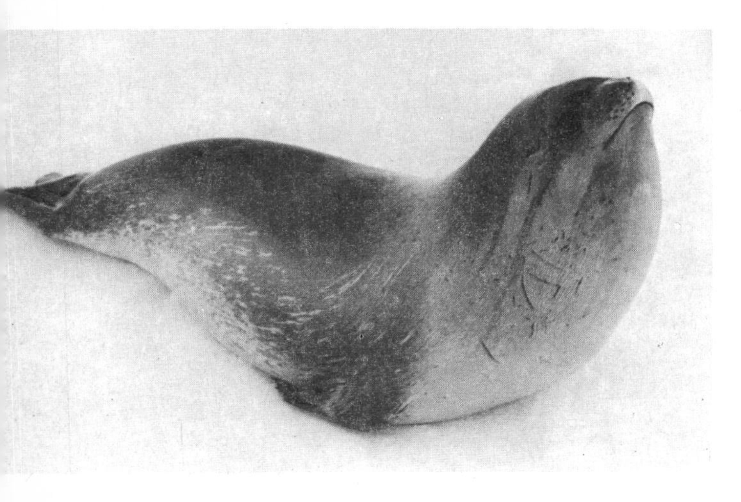

The rarest of Antarctic seals, the Ross seal, lives on and around pack ice far out at sea, where it eats squid and fish.

which congregate in well-defined areas around Antarctica. Blue whales, fin whales, and right whales also feed on the krill and gather around Antarctica at this time. When the supplies of krill dwindle later in the year all these whalebone whales migrate northward up the coasts of Africa, South America, Australia, and New Zealand. During the winter they live mainly off their blubber, because krill are not found in these warmer waters. While in these warm areas the whale cows give birth to young. The calves feed on the cows' rich milk and are soon strong enough to make the long journey south to the Antarctic, arriving in time for the new summer crop of krill.

Although the great whales are not strictly polar animals, because they travel annually

between the Antarctic and temperate areas, they have many adaptations to help them to survive in the cold polar seas. Although they are very large animals, they have a relatively small area of skin through which heat can be lost. Their beautifully streamlined bodies, perfectly shaped for active swimming, are very compact and thus conserve the maximum amount of heat in cold water.

Whales are mammals and are therefore warm-blooded. Because they can maintain a constant body temperature all the time, they can be just as active in cold polar seas as in temperate waters. Instead of having an insulating layer of fur, whales are almost hairless, but they do have a layer of blubber up to one foot thick beneath the skin, serving the dual purpose of insulation and streamlining.

Seals are less completely aquatic mammals than whales because they often haul themselves out of the water to bask in the sun, and in the summer to breed. For most species of seals found in the Antarctic their home is not the frozen land but the permanent floes of the pack ice.

The solitary Ross seal, the leopard seal, and the more sociable crab-eater seal live far out at sea all the year round. All three species spend their time alternately in the water and on the ice. Although they all feed in the same waters, each kind specializes in a different type of food. The Ross seal has hooked needlelike teeth for seizing squid and fish. The crab-eaters, despite their name, strain krill from the water through special triple-pointed teeth, and grind the food up by means of gravel and pebbles in their stomachs. And the leopard seal hunts alone for its warm-blooded prey.

The Weddell seal lives in the coastal waters around Antarctica. During the winter the air temperature is so low that the seal can survive only by staying in the water, and it may remain beneath the ice for days or even weeks. To breathe, the seal scrapes holes through the ice, sawing with its teeth while moving its head from side to side. This is very hard work and can wear down the seal's teeth. When an old seal is unable to keep its breathing holes open in this way it is in danger of suffocating unless it can find holes made by others.

A seal's shape, like that of a whale, is its chief adaptation for life in the polar oceans. It, too, has a thick layer of blubber beneath the short waterproof fur. The small vessels carrying blood through the blubber to the skin can be closed off during the winter so that less blood travels to the surface and the seal's heat loss is greatly reduced. Even with these adaptations, seals may freeze to death on the ice if they cannot get back into the sea.

The walrus is a typical polar mammal in several respects. The Atlantic and Pacific races of walruses belong to a single species distributed in shallow waters right around the Arctic region. Their outlandish appearance is the result of an almost perfect adaptation to Arctic conditions. The lumbering body, which can measure 11 feet in length and weigh 3000 pounds, has a large volume but a relatively small surface area. This reduces the loss of valuable heat by radiation from the animal's skin. Like the whales and the seals, the walrus has a protective layer of blubber several inches thick under the skin. Compared with its closest relatives, the fur seals and sea lions, which live farther south, the walrus has the heaviest but most compact body and the smallest flippers.

Walruses are bottom-feeders. They dive to depths of about 150 or 200 feet to feed on clams, mussels, and snails, which they dig out of the mud or pry off rocks with their tusks. The long, stiff vibrissae on the snout are used for locating food by touch in water where visibility is poor. The tusks, which are enormously elongated upper canine teeth, continue to grow throughout life, reaching a length of three feet in old bulls. They are also used for fighting over cows, for self-defense, for hauling the walrus out of the sea onto ice, and for keeping open breathing holes in the pack ice. Although the walrus may appear awkward on land, humping itself along on its short flippers, it can move over short distances as fast as a running man and is a formidable enemy when roused.

To most people, the first animal that comes to mind when thinking of the Arctic is not the walrus, but the polar bear. The polar bear's white fur camouflages it against the snow and ice when it is hunting, and the fur is also an excellent insulation against the Arctic cold. Beneath the coarse six-inch-long guard hairs lies a thick layer of soft underfur, which traps air next to the skin. Moreover, although polar bears usually hunt on the ice, they often swim across long stretches of water, and their oily outer fur is water-repellent, so that a quick shake leaves it almost dry after a swim. Even the feet of a polar bear, which are wide and flat for swimming, have fur on the underside, which helps to give the bear a good grip on slippery ice.

The polar bear is more completely carnivorous than any other kind of bear because for most of the year no plant food can be found in the barren Arctic snowfields. It lives mainly on seals, which it kills with a powerful blow of the forepaw as they come up to breathe at cracks in the ice. Polar bears also eat fish, carrion, seabirds, and small mammals, and occasionally lichens and berries in the summer.

The most characteristic animal of Antarctica is the penguin, which, in spite of not being able to fly, survives because there are no large predators to hunt it on land. The largest of all penguins is the emperor penguin. It copes with the most severe conditions any bird has to face. Compared with the smaller Adélie penguin, with which it often forms large rookeries, it has relatively small wings and feet to reduce its surface area. A penguin's feathers are small and lie flat, trapping a layer of insulating air at their fluffy bases. The feathers overlap and are covered with oil from the bird's preen glands. Beneath the skin lies a layer of blubber, so that an adult penguin is well insulated. Walruses and emperor penguins tend to overheat, which is one reason you seldom see them in zoos in the warmer parts of

Huge herds of walruses sunbathe on rocky islands during the Arctic summer. A thick layer of blubber under the skin of its bulky body helps the walrus to withstand intense winter cold, and the fat serves as a food reserve. The ivory tusks are used, among other things, for self-defense and for scraping mussels off rocks.

the world. Penguins are able to lose a little surplus heat from the skin by fluffing out the close-lying feathers and spreading their flippers to allow the air to cool them.

Young penguins have to grow up quickly so that they are strong enough to survive their first hard winter. This is also true of all polar mammals—bears, seals, walruses, and whales. A constant body temperature and a rich supply of milk from their mothers help to speed their growth. All these animals have a slow rate of reproduction, however. Emperor penguins and most seals produce only one offspring in each breeding season, and polar bears and whales breed every two years on average. This means that losses in the total numbers are made up slowly. Because the delicate balance between survival and death under harsh polar conditions is easily upset, these animals are particularly vulnerable to changes in external conditions, including interference by man.

For the animals adapted to cope with the very severe conditions of polar winters, the Arctic and Antarctic provide enough space and food for large populations, as the crowded colonies of penguins and fur seals, and the great schools of cold-water fish show. The warmer waters and, to us, the easier living conditions of the tropics might be expected to support even larger numbers of animals, but in fact this is not so. True, the number of different varieties of animals in a given area is much greater in tropical than in polar seas, but there are relatively few individuals of each kind. Apart from the coral animals, whose existence depends on the communal life of

Polar bears' thick fur allows them to sleep in the open all the year round, but females den up in winter before giving birth to their cubs. Young bears stay with their mothers for about 18 months, learning how to survive, after which they fend for themselves. Adult bears congregate only during the mating season.

A pair of emperor penguins attend to the needs of their newly hatched chick. Emperors are the largest of penguins and live farthest south, on the Antarctic shores. Eggs are laid in midwinter and incubated by the male, but when a chick hatches, it is cared for mainly by the female, which keeps it warm beneath a flap of skin over her feet.

Each of these daisylike heads of green soft coral is a single animal, with a ring of tentacles that sting and paralyze the tiny planktonic organisms that serve to feed the entire colony.

the coral colony, most tropical species are rather scattered. Reproduction and growth in the tropics are not restricted to certain seasons, and generations of mobile larvae effectively disperse the crowds of offspring.

The tropical oceans form a permanent belt of warm water round the equator, and a typical feature of these waters is the coral reef. The corals that form these structures are tiny colonial animals related to the sea anemones and jellyfish. Each coral animal, or polyp, builds a hard, chalky skeleton around itself, and the skeletons join together to form a mass known as the coral head. As the founders of a colony die, their compounded skeletons support the later generations of polyps. Coral animals catch minute planktonic organisms in their tentacles. During the daytime they are withdrawn inside their strongholds, but at night, as the plankton population swims up to the surface, the tentacles come into action, stinging and capturing food for the whole colony.

The main conditions necessary for the growth of corals are a minimum water temperature of 70°F and a maximum depth of 150 feet. Shallow clear water is necessary because many corals are dependent on minute plants for part of their diet, and these plants must have sufficient light. The living coral animals are small and soft, and they vary little between species, but the fantastic shapes of the coral heads they build have given rise to such names as brain coral, stag's horn, sea fan, and sea feather. These shapes are the result of different patterns of growth in the colonies, with new polyps budding off from the parents in characteristic ways. The shape of a mass of coral also depends on water conditions. In still, deep water, the polyps can build unhindered in any direction, forming many branches. In shallow water stirred by waves the outer polyps are constantly being damaged and so an irregular solid mass is formed.

There are three main types of coral reef: fringing reefs close to the shores of islands or continents; barrier reefs much farther out and separated from the shore by a deep channel; and atolls, which are independent rings of coral not associated with land. All three types are constantly changing. They are added to by coral growth, and subtracted from by the action of waves and such coral-eating animals as the parrot fish and crown-of-thorns starfish. Under ideal conditions, coral reefs seem to grow at a rate of between one and two inches a year; and in some places boring into them has revealed that the coral is thousands of feet thick. As we have seen, coral can grow actively only in relatively shallow water, and for centuries biologists and geographers have been trying to discover how these great piles of chalky material have been formed.

It is probable that coral can begin to grow along any shore where the conditions are right. First a fringing reef is formed, between the low-water mark and a depth of about 150 feet. On the outer (seaward) side the slope of the reef is steep, and boulders of coral are broken off by the waves and hurled up onto the reef, forming a raised outer rim. As a result either of the land slowly subsiding or of the sea level slowly rising, the coral is able to grow ever upward with the passage of time—and so, gradually, great thicknesses of calcareous material build up. If the reef grows around an island it slowly comes to lie farther from the shore, becoming a barrier reef. Eventually subsidence or rising sea level may cause the island to be completely submerged, and the reef will remain as a ring (or atoll) surrounding a central lagoon connected with the sea by channels through the coral.

The conditions that lead to the building of reefs in shallow warm water allow a dazzling collection of other animals to live among them. Many brilliantly colored fish, including the damselfish, the clownfish, and the butterfly fish, dart among the coral branches. These fish eat a varied diet, and some, such as the long-nosed butterfly fish (*Forcipiger*), have long snouts for probing into cracks in the coral in search of worms and small crustaceans. Others have small mouths and little sharp teeth for nipping off coral polyps at their base.

The most outstanding feature of these reef fish is the brilliance and variety of their colors and patterns, and these are further enhanced by their darting activity. The many species of butterfly fish, damselfish, clownfish, and surgeon fish use their conspicuous colors to flaunt possession of their territories, rather as we might use flags. Each fish is immediately recognized by others of its kind, and so in the clear waters around coral reefs they can avoid each other's living space. In a reef there are unlimited different corners and crannies to live in, and this is why so many patterns have evolved among these fish.

The vast number of crevices and caves provided by these corals serve as a shelter for scavenging crabs, sea urchins, and many kinds of fish. And filter-feeding sponges, sea squirts, and tube-building worms cling to the surface of dead coral heads. All of these creatures are part of a typical reef community.

The only pleasant feature of the moray eel is its patterned skin. This vicious predator, which can grow to a length of eight feet, hides among rocks and coral reefs, ready to snatch any animal that swims within reach of its sharp teeth.

Another reef resident is the ferocious moray eel, whose doglike jaws are lined with vicious teeth. Like all eels, the moray has no protective scales, and so, rather than expose itself to danger from other predators, it lurks in deep crevices and grabs unsuspecting fish as they swim past.

Some animals have become adapted to feed directly on the coral itself. Parrot fish have a pair of heavy, thick teeth in each jaw; the toothed jaws function like a parrot's beak, enabling the fish to bite and crush the hardest coral skeleton before digesting the soft animals within. These fish, whose great range of color constantly baffles the zoologists who try to classify them, sleep on the seabed at night, encased in a mucous sheath that they secrete. The mucous sheath probably functions both to anchor them down and to protect them from the sharply pointed coral masses.

Another notorious coral-eater is the large crown-of-thorns starfish, which has been responsible for killing off acres of coral in Australia's Great Barrier Reef, and in the Red Sea. This starfish feeds on living polyps by pouring its stomach juices over them. The starfish do enormously more damage than the many boring clams and sponges, for these creatures feed mainly on old dying coral heads. When a crown-of-thorns starfish moves on after a meal it leaves a devastated patch of coral, which is unable to rebuild itself.

The crown-of-thorns starfish has undergone a population explosion during the last five to 10 years, possibly because one of its few enemies, the beautiful triton snail, is becoming rare due to over-collection. The best hope for the future of the corals rests in the fact that two types of reef fish also prey on the crown-of-thorns: the little pufferfish nips off the starfish's spines, from which it is protected by its own armor; and the triggerfish has an ingenious way of picking up a starfish by one leg, flipping it onto its back, and then attacking the soft vulnerable underparts.

Because of the almost total lack of algae on a coral reef, the majority of animals living there are either carnivores or scavengers. But along tropical coasts and in shallow water at the mouths of rivers there is sometimes quite a rich flora of seaweeds and other plants. In these waters can be found the sea cows. These bulky mammals, which never leave the water, seem superficially to resemble both the whales and the seals in some aspects of their lives, but they are in fact related to neither. Their nearest living relatives are land-dwellers—the elephants and

the hyraxes—and it is probable that these apparently very dissimilar creatures all shared a common ancestor about 60 million years ago. Since the huge Steller's sea cow became extinct 200 years ago, a mere 30 years after its first discovery, the only living sea cows have been the dugongs and manatees.

Dugongs live along the coasts of the Indian Ocean, the Red Sea, and the western Pacific Ocean, browsing at night on marine flowering plants such as eelgrass and waterweeds. They have paddlelike forelimbs, but no hind limbs and are quite unable to move on land. Vertical movements of the broad, flat tail fin, with its two distinct flukes, push the animal slowly through the water. Like other animals adapted to a marine habitat, the dugong has nostrils on top of its snout, and they can be tightly closed during a dive, which usually lasts about 10 minutes. Sea cows have the densest bones of all known mammals, which allow them to stay near the sea floor to feed with little effort.

The manatees live in the tropical estuaries around the Caribbean, Central America, and West Africa. They sometimes venture into fresh water, and a group of them has been seen in the river in the center of Miami, Florida. Like the dugong, they are completely herbivorous, and in Florida they play a very useful role in eating the rapidly growing water hyacinth, which blocks waterways if not kept under control.

In temperate and tropical regions, the great majority of reptiles are terrestrial or freshwater animals. However, a small number live in the tropical seas, and deserve a mention here. None is completely adapted to a marine life and, like all reptiles, they must come to the surface to breathe air, and return to land to lay their eggs.

In a coral reef typical of tropical seas (left), brightly colored small fish, whose stripes are probably a form of camouflage, swim among the varied shapes of coral heads—among them, branched stag's horn coral and folded yellow soft coral. A link in a coral-reef food chain is pictured above: the pufferfish is biting off the spines of a crown-of-thorns starfish, which in its turn has been feeding on living coral polyps. Such starfish do great damage to reefs, for they have large appetites and few enemies.

Unlike amphibians, which have a naked, permeable skin, marine reptiles have no problem with salt and water balance because of their scaly, impervious skins.

The most numerous sea reptiles, although not the best-known, are the sea snakes of the Indian and Pacific oceans. They have long flexible bodies and their tails are flattened from side to side, and so they are able to swim well, using the sinuous movements typical of snakes. Sea snakes belong to the same family as the cobras, and have a powerful venom, which is chiefly used for killing the fish that they eat.

Turtles are the most familiar group of sea reptiles. There are several kinds, spread throughout the tropical oceans of the world. Most of them eat a mixed diet of fish, crustaceans, mollusks, and seaweeds, but they have individual preferences, too. The loggerhead, for instance, specializes in crushing mollusks with its powerful beaklike jaws, and the green turtle (the unfortunate source of turtle soup) is herbivorous, feeding primarily on seaweed. The most striking thing about the marine turtles is their great size compared with that of most land turtles: the marine turtles grow to lengths of up to eight feet and may weigh up to a ton. With the sea to support this weight, and limbs modified as paddles, they are good swimmers.

Around the shores of the Galápagos Islands, off the Pacific coast of Ecuador, live the only seagoing lizards, the marine iguanas. At one time these animals, which can be up to five feet long, were found swarming in great crowds over the rocks; but as a result of the ravages of dogs and pigs brought to the islands by sailors, they are now much less common. The marine iguana's brownish black skin does a good job of camouflaging it on the dark rocks. When alarmed, however, it normally takes to the water. Like all iguanas it is herbivorous, and seaweeds, for which it dives, make up its entire diet. The waters around the Galápagos Islands are cooled by the north-flowing Peru Current from the Antarctic, and so the marine iguanas must warm up on the rocks after a dive. If they did not come out of the water, their body temperature would drop too low for them to continue active swimming. After lying about on the hot rocks for a while, they become overheated and must return to the water. So their lives are a continual battle to maintain a moderate temperature.

This survey of animal life in polar and tropical seas has given some account of the rich and varied collections of animals to be found in these waters, despite the extreme conditions. However, there are some parts of the oceans that support only a small number of organisms. It is interesting to compare such an "ocean desert"—for example, the 2-million-square-mile Sargasso Sea in the middle of the Atlantic—with a land desert such as the Sahara. The popular idea of a desert is of a vast empty area, too hot and dry to support life. In fact, though, many living things

A group of manatees, which live in the warm waters of tropical river mouths in Central America and West Africa. The manatees and dugongs are the only living sea cows—strictly herbivorous marine mammals, which browse on flowering plants and are in no way related to whales and seals.

can be found in even the hottest and driest places. But it is also a fact that only the best-adapted plants and animals can live under such extreme conditions, and so the Sahara supports a very limited number of species.

In the Sargasso Sea, conditions would seem, superficially, to be much more favorable for living things, with no lack of water, no extremes of temperature, and no strong winds or violent storms. But the very tranquillity of this mid-ocean area is the main reason for its limited animal and plant life. Very little mixing of water occurs; and so, although the salinity is high because of evaporation, the supply of nutrient minerals normally carried by currents is poor. Without these minerals there is little plankton to feed small pelagic animals, and this in turn means a poor supply of food for the larger fish, birds, and mammals. Because the Sargasso Sea lies at the center of the North Atlantic Current system, new plants and animals are carried into it only occasionally, and by chance. So the evolution of varied forms as a result of interbreeding is a slow process here.

The characteristic brown seaweed of the Sargasso Sea, which has given the area its name, is covered with small gas-filled bladders that look like grapes (*sarga* is Portuguese for "grape"). This weed, torn from the shores of the Caribbean, accumulates in the Sargasso Sea as a result of the circulating currents. In patches it forms a thick layer on the water's surface, sometimes even obstructing the passage of ships. The blanket of weed naturally prevents light from

Most reptiles are either land or fresh-water creatures. Among the few that live in the sea are the marine iguanas of the Galápagos Islands (above) and the banded sea snakes of the Indian and Pacific oceans (left). The iguanas, which eat nothing but seaweed, maintain a moderate body temperature by alternately warming up on sun-baked rocks and cooling off in the sea. The sea snake, a very good swimmer, uses its powerful venom to kill the fish that it eats.

Marine turtles are also fully adapted to life in the sea, although they lay their eggs on land, as do most reptiles. After her laborious nocturnal stint of digging a sand pit in which to conceal up to 100 eggs (left), this green turtle crawls back to the sea at dawn, leaving a tanklike track down the beach (above). Scavenging wild pigs and dogs often follow the tracks of the turtle to find the freshly laid eggs, which make a tasty meal.

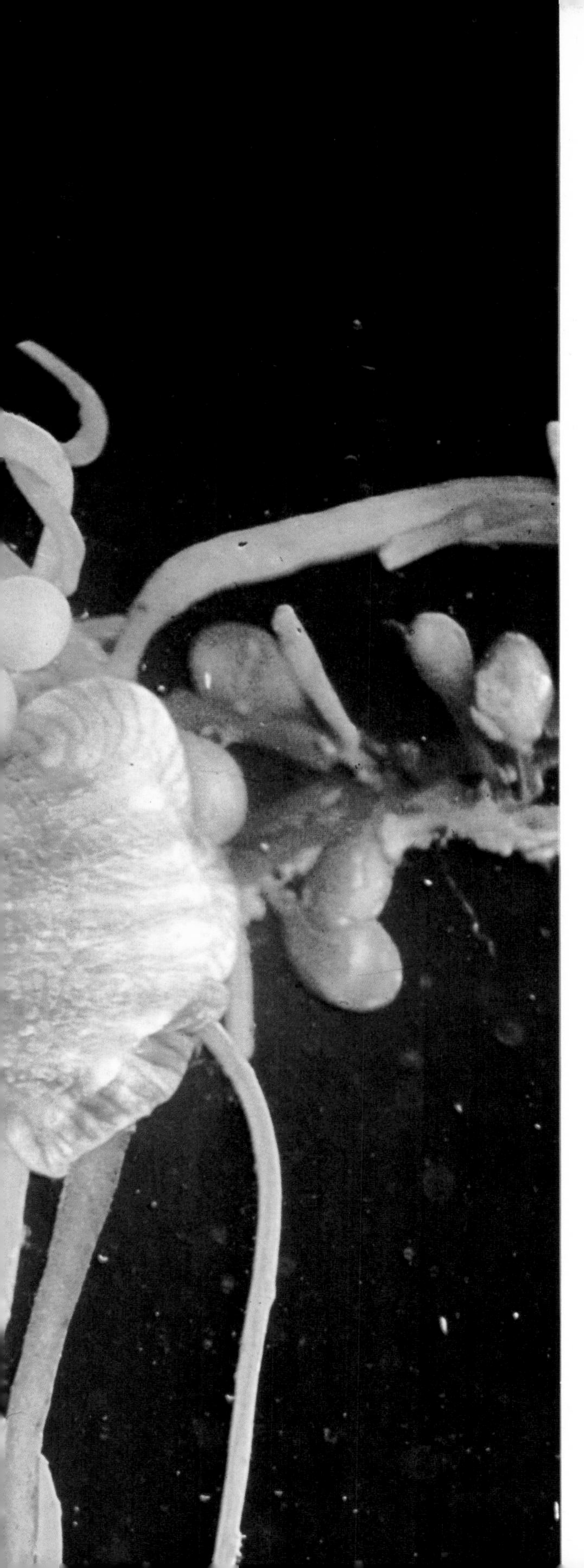

penetrating beneath the surface, and this, with the lack of minerals, limits the growth of the phytoplankton. Once the sargassum weed is free-floating it cannot reproduce; however, it breaks into pieces and continues growing; it is practically immortal once it has reached the area of calm.

Transported to the Sargasso Sea on the fronds of the weed are a number of sedentary shore animals such as hydroids and barnacles, and also small shrimps with hooked legs, well adapted for hanging onto the strands of weed. Then there is the remarkable sargassum fish, which has evolved such excellent camouflage, complete with false air bladders, that it looks like a clump of the weed among which it lives. A few flying fish swim in the upper layer here, and in their nests of weed they lay eggs that also look very much like the plant's bladders.

Apart from the legends of ships becoming trapped forever in the masses of sargassum weed, the Sargasso Sea is well known also as the spawning ground of eels. Nobody ever sees the eels' eggs, but their smallest larvae are found about 600 miles east of Bermuda. They must be able to swim actively to leave the currentless area and reach the Gulf Stream, which carries them to the rivers of Europe and eastern North America. In some mysterious way, the eels find their way back as adults, six or seven years later, to the very same part of the ocean in order to spawn.

We have seen that the oceans can be divided into different ecological zones almost as clearly as the earth's continents, although no definite boundaries separate the cold, temperate, and warm regions. In the sea the zones depend upon the movements of currents and upon the depth of water, rather than upon latitude or climate. Animal life in polar and tropical seas illustrates the two extremes of adaptation. But marine animals are not restricted by physical boundaries and there are many fish and whales that are widely distributed throughout the world. In contrast, the boundary between sea and land is distinct and on the fringes of the oceans there are relatively few animals that can move freely between one habitat and the other.

Among the tassels and air bladders of surrounding weeds, a sargassum fish and a sea anemone are beautifully camouflaged. Anemones are sedentary animals, and this one probably reached the calm waters of the Sargasso Sea attached to a weed carried along by one of the North Atlantic currents.

The Fringes of the Sea

The *littoral*, or intertidal, zone, where land and sea meet, is a well-studied habitat. The majority of animals and plants that live there come originally from the sea rather than from the land and show a range of adaptations by which they overcome the problem of alternate exposure to the air and submersion by the sea.

The limits of the littoral zone are clearly defined. It lies between extreme high-tide level and extreme low-tide level. In most parts of the world the maximum, or spring, tides occur twice in each lunar month, at new moon and full moon. At these times the sun, moon, and earth are almost directly in line, and the combined attraction of the sun's and moon's gravities causes a pile-up of the oceans on the side of the earth nearest to them. When the moon is in its first or last quarter, the sun, moon, and earth are at right angles and so the pull of the sun counteracts the pull of the moon, resulting in the smaller neap tides. On the side of the earth farthest from the gravitational influence of the sun and moon, the land part of the earth is closer to the moon than the water. The land is, therefore, pulled closer to the moon than the more distant water, which is "left behind," causing a second high tide to occur. Because the earth rotates on its axis once in approximately every 24 hours, most parts of the earth experience two high tides every day.

As a result of such other influences as the distances of the sun and moon from the earth, their positions north or south of the equator, and the rocking movement that occurs in every body of water, the study and prediction of tides is very complex. Some landlocked seas—the Mediterranean is probably the best-known example—are practically tideless, because they are isolated from the main bodies of water. Other places may experience extreme tidal changes of 50 feet or more, as in the Bay of Fundy where the record tide of 53½ feet was measured. Farther down the coast of eastern North America, around Cape Cod and Nantucket Island, the tide range is only one or two feet. Not all parts of the world have two tides a day; in the Gulf of Mexico, for example, the tide rises and falls only once in every 24 hours.

Many birds live in and around the area where sea and land meet, but few of them can be said to live literally on the edge, as gannets do. Crowded communities of the large birds, such as this one on Bass Rock in Scotland's Firth of Forth, contain thousands of pairs of noisy nesting cliff-dwellers. They make spectacular dives into the water to catch and eat fish.

The tide range between high and low water is measured vertically. The same tidal movement may cover and uncover a variable area of shore, depending on the physical structure and the slope of the coastline. In some places great stretches of beach are exposed at low tide, but in others only a few extra inches of rock appear. The extent and timing of tidal movements thus varies considerably, as does the nature of the sea coast itself. In their struggle for survival, however, all the animals and plants found on the shore face the same difficulties: the considerable changes in humidity and temperature, and the often violent movements of the water around them.

On a rocky shore, organisms at the foot of cliffs live in a clearly visible pattern of vertical zones. The splash zone, just above high water, is often marked by a band of gray, black, and yellow lichens, which are simple terrestrial plants in which fungi and algae live in close association. These are never completely submerged but must survive higher concentrations of salt than terrestrial flowering plants such as thrift and sea lavender, which grow farther up the cliffs.

The typical plants of the upper intertidal zone of a rocky shore are the green seaweeds *Enteromorpha* and *Ulva*. Both are particularly abundant on rocks where fresh water is available and in high-level pools. *Enteromorpha* has long, tubular, gas-filled fronds, which become dried and bleached to white cords by the late summer. *Ulva* has irregular broad fronds, which spread over the rocks in a green sheet. Among these green weeds live a number of animals that can breathe in either air or sea water—sea slaters, beach fleas (or sandhoppers), bristletails, and the small periwinkle.

In the middle region of the shore the thin covering of green algae gives way to thick mats of brown seaweed called *wrack*. Progressing down the shore toward the sea we can find channeled wrack, flat wrack, and toothed wrack.

Above: a splendidly camouflaged spider crab lying among the multicolored seaweeds in a rock pool, which is a permanent pool of sea water formed in a small depression on a rocky shore. Weeds and sponges hooked on the crab's back, mask it from its prey.

The common green seaweed (Ulva) *known as sea lettuce is found on most rocky shores of the North Atlantic Ocean. Its thin, flat fronds are especially in evidence wherever fresh water occurs on the upper shore, and also in rock pools, where it is the food of browsing mollusks. Some human beings eat it, too.*

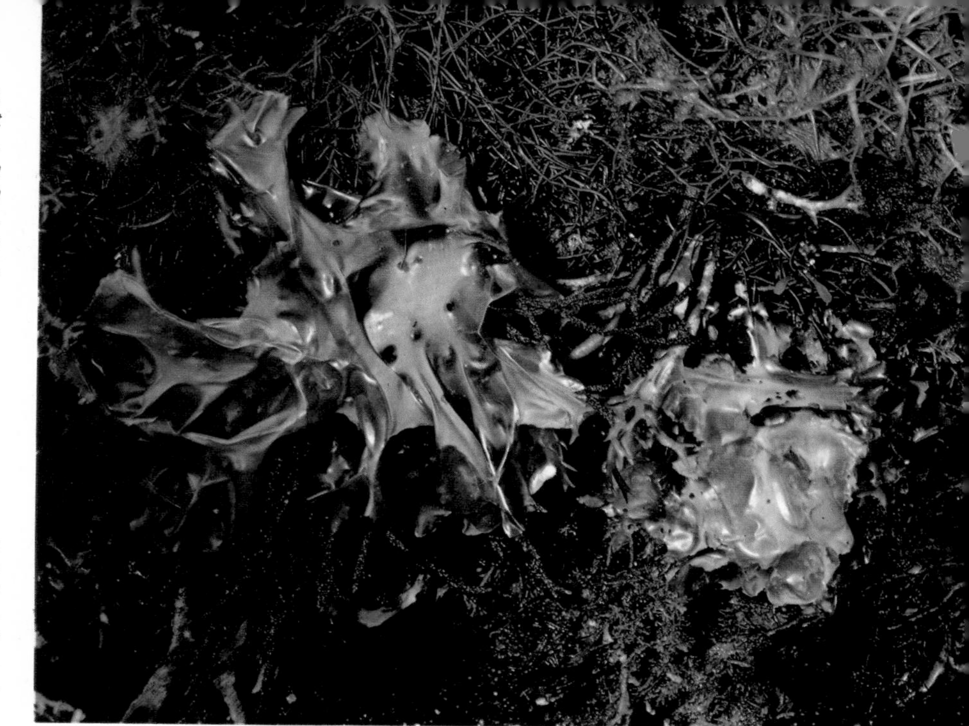

Below: typical rocky shoreline, on Mount Desert Island, Maine. Brown seaweeds that cover the rocks of the intertidal zone are alternately exposed and submerged by the tides. Large numbers of animals live here, both in the sea water of the rock pools and in the permanently moist tangle of fronds.

A Simple Food Web on a Rocky Shore

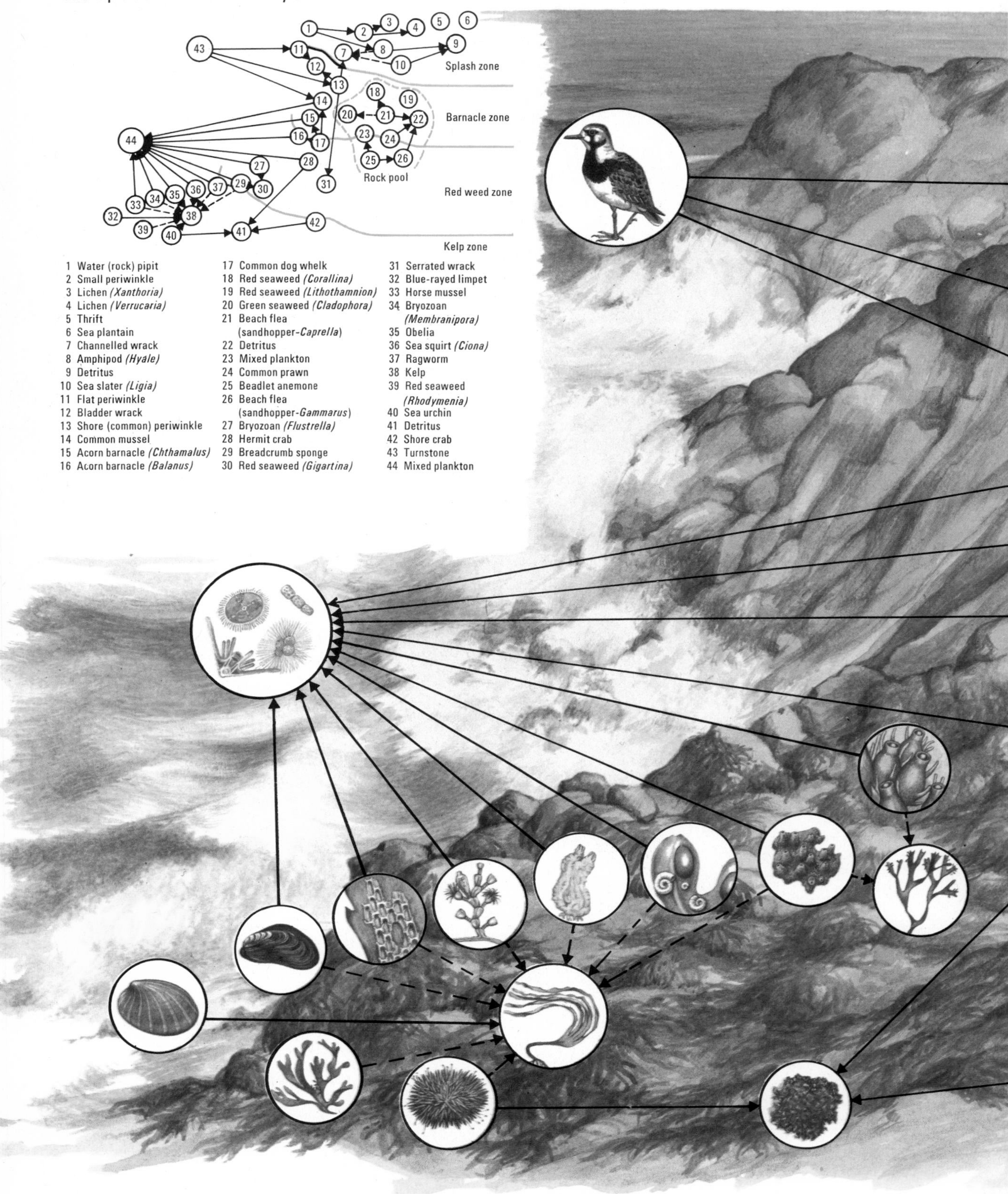

1 Water (rock) pipit
2 Small periwinkle
3 Lichen *(Xanthoria)*
4 Lichen *(Verrucaria)*
5 Thrift
6 Sea plantain
7 Channelled wrack
8 Amphipod *(Hyale)*
9 Detritus
10 Sea slater *(Ligia)*
11 Flat periwinkle
12 Bladder wrack
13 Shore (common) periwinkle
14 Common mussel
15 Acorn barnacle *(Chthamalus)*
16 Acorn barnacle *(Balanus)*
17 Common dog whelk
18 Red seaweed *(Corallina)*
19 Red seaweed *(Lithothamnion)*
20 Green seaweed *(Cladophora)*
21 Beach flea (sandhopper-*Caprella*)
22 Detritus
23 Mixed plankton
24 Common prawn
25 Beadlet anemone
26 Beach flea (sandhopper-*Gammarus*)
27 Bryozoan *(Flustrella)*
28 Hermit crab
29 Breadcrumb sponge
30 Red seaweed *(Gigartina)*
31 Serrated wrack
32 Blue-rayed limpet
33 Horse mussel
34 Bryozoan *(Membranipora)*
35 Obelia
36 Sea squirt *(Ciona)*
37 Ragworm
38 Kelp
39 Red seaweed *(Rhodymenia)*
40 Sea urchin
41 Detritus
42 Shore crab
43 Turnstone
44 Mixed plankton

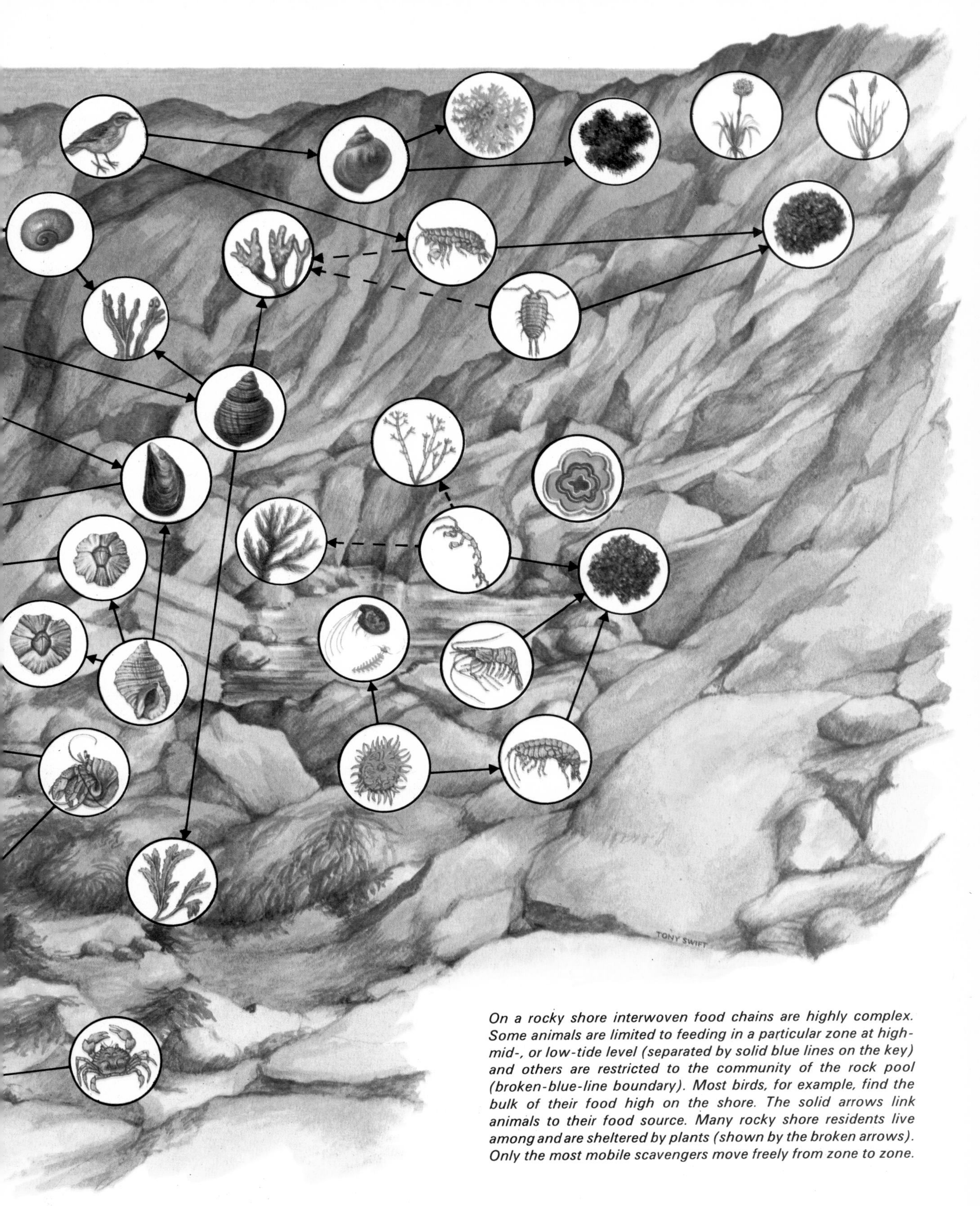

On a rocky shore interwoven food chains are highly complex. Some animals are limited to feeding in a particular zone at high-mid-, or low-tide level (separated by solid blue lines on the key) and others are restricted to the community of the rock pool (broken-blue-line boundary). Most birds, for example, find the bulk of their food high on the shore. The solid arrows link animals to their food source. Many rocky shore residents live among and are sheltered by plants (shown by the broken arrows). Only the most mobile scavengers move freely from zone to zone.

Among the fronds live many familiar animals: acorn barnacles and limpets clinging to the rocks, rough periwinkles browsing on the weeds, and predatory dog whelks. All these animals can seal themselves inside their shells to avoid drying up when the tide is out. Farther down the shore there are common and flat periwinkles, mussels, chitons, top shells, tube worms, shore crabs, and sponges.

Just above low water in the area uncovered only at low tide, the dominant plant is the brown tangleweed, with its flat, straplike fronds. Even when uncovered this tangled mass of weed holds enough moisture to enable animals that normally live permanently in water to survive until the tide comes in again. Sea anemones, corals, brittle and common starfish, edible sea urchins and crabs, prawns and lobsters, sea slugs, cowries, and various small fish may all be found among the weeds of this lower shore zone.

Because the shore organisms came originally from the sea, it is not surprising that the farther up the shore we go, the fewer individuals and the smaller the number of different varieties we find. Few modifications for breathing, movement, or protection are needed by animals that may be uncovered for less than an hour a day. However, at the other end of the scale—the upper intertidal zone, where the shore is exposed for most of the time—most of the organisms are semiterrestrial, such as *Enteromorpha*, and the small periwinkle, which breathes by way of a simple lung.

On a typical rocky shore, small depressions in the rocks, which permanently hold some sea water, create the fascinating miniature worlds of the rock pools. Although they can be considered as natural aquaria with balanced communities of plants and animals, we must remember that rock pools change rapidly and widely, both in their conditions and in their inhabitants. When the tide goes out, a pool high on the shore will warm up quickly in the sun, and the salinity rises because of evaporation. During the daytime in such a pool oxygen from the photosynthesizing weeds accumulates. At night less oxygen is produced, but to balance this the respiration rate of the cold-blooded animals in the pool falls as the temperature drops. These variations are exaggerated in warm water and less marked in colder water, but they are considerably greater than the changes that occur in the open sea.

Most rock pools are not large enough to hold the long fronds of brown algae, although young plants are often present. Many small green, red, and blue-green weeds that are usually found below the low-tide mark can grow in these pockets of sea water, once they have been carried in by the waves. Rock pools, too, are the best places to see expanded sea anemones, which usually appear only as dark, gelatinous blobs among weeds on rocks uncovered at low tide. Brightly colored beadlet and dahlia anemones and the aptly named snake-locks anemone trap animals in their waving tentacles. Jewel anemones and cup corals spangle the sides of pools with their flowerlike heads.

A variety of crustaceans scavenge for food dropped by the anemones. Prawns normally spend the cold months in the security of offshore waters, but in early summer they move up into the rock pools. It is during the summer months that lovers of seafood become regular hunters along the shore. Often transparent or camouflaged by spots and stripes, the prawns reveal themselves only by darting backward across the pool when disturbed. They have well-developed eyes and two pairs of feelers for locating their food, which they pick up in pincer-tipped legs. Prawns mate and lay their eggs in the pools during the summer.

The mollusks found in rock pools include many that are more typical of the exposed parts of the shore, such as winkles, limpets, and mussels. But animals normally found only in rock pools are the shell-less sea slugs, with a variety of colorful feathery gills and frills fluttering around their

A common animal of rocky shores is the limpet, which glides over rocks and clears them of weeds, like a grazing sheep. Clinging tightly by its muscular foot, the limpet (seen here from below) can withstand constant battering by waves.

Like limpets, the brown seaweed kelp is perfectly adapted for life in the wave-buffeted intertidal zone. Its narrow, straplike fronds and tubular stems resist the tearing force of the water by swaying back and forth with each wave. Special organs called holdfasts *become so firmly attached to their rock base that they can be pried off only with great difficulty, even when the main fronds are dead and dry.*

bodies. One of the largest is the sea lemon, whose tough yellow body is well hidden among the rock-encrusting sponges on which it feeds.

Among other, more active inhabitants of rock pools are small fish such as bullheads, shannies, and sucker fish. The body shape of these fish is flattened, either from side to side, or from top to bottom, so that they can squeeze themselves into crevices in the rocks to avoid predators, desiccation, and wave action.

One of the main hazards to survival for animals and plants living permanently on a rocky shore, whether in pools or on exposed rocks, is the constant motion of the waves. Unless they are firmly attached to something solid they are in danger of being swept away on the tide or battered to death by the force of the water. One way of overcoming this problem is to live on the more sheltered surfaces of the rocks, in crevices and on the leeward side. This is the habit of the dog whelk and periwinkles, which may often be found in rows jammed into cracks between rocks. Like all the snail family, these mollusks cling by means of their muscular feet, aided by a secretion of sticky mucus. Their coiled shells have the mechanical advantage of being able to withstand violent water movements by rolling with the current while still attached to the rocks.

On the most exposed rock faces in the intertidal zone live the limpets. After a planktonic larval stage, the young limpet settles and becomes a browser on seaweeds. The adult clings to the rock by the suction of its powerful muscular foot, which forms a vacuum on the smooth surface

The boldly patterned black-and-white plumage of the oyster catcher—a common shore bird—helps to conceal it against a rocky background. Its orange-red flattened bill is used by the bird as a chisel to scrape limpets off rocks and to pry open the shells of bivalve mollusks, which form part of its diet.

beneath. Anyone who has tried to dislodge a limpet will know that the best way to do this is by a sudden sharp blow. If the animal is warned by a gentle tap, it pulls its shell down tight onto the rock and becomes almost impossible to remove. This is because there is a perfect contact between the shell and the rock in one special place to which the limpet always returns by a homing instinct. On a hard rock the animal's shell is ground to fit, but on a soft rock the edge of the shell cuts a ring-shaped groove in the rock itself. The conical shell, which grows from the base, is shaped to withstand the full battering force of the waves. Surprisingly, limpets found on the most exposed rocks have higher, narrower shells than those in more sheltered pools, where the downward and inward pull of the muscular foot on the growing region of the shell is less marked.

Competing with the limpets for the title of "best-adapted animal" on exposed rocky shores are the acorn barnacles. Found by the million encrusting the rocks from high to low water, these strange crustaceans live very specialized lives. They pass through a free-living larval stage in early spring and do not begin their sessile adult lives until they have explored the rocks for a suitable spot on which to settle. Barnacles require rough surfaces and some shade, and they cannot settle if the tidal movement is too great. Once the barnacle larva has settled, it grows a ring of six hard, limy plates, which it cements to the rock by means of the remains of its sensory feelers. These fixed plates surround four smaller movable plates. At low tide internal muscles draw all the shell plates together, and the animal is perfectly protected against desiccation and predators. When covered by the incoming tide, the barnacle opens and puts out its feathery legs to sweep the water for plankton.

Another method of attachment especially suitable for an exposed rocky habitat is that used by the common mussel. In this bivalve mollusk, the foot muscles must be used to hold the two valves of the shell tightly together, and so each animal grows a tuft of strong fibrous threads ending in round disks for attachment to the rock surface. These *byssus* threads are formed from a sticky liquid that hardens on contact with air or water. They radiate out from the mussel like guy-ropes on a tent, so that in whichever direction the shell is buffeted by the sea, some of the threads are able to take the strain. The shape of a mussel, with one end narrow and pointed and the other rounded, also helps it to withstand wave action.

Starfish and sea urchins are more mobile than mussels. But the tube feet on which they move slowly about in search of food are also effective for holding onto rocks when necessary. Numbers of tiny suckers arranged in rows on the underside of the starfish's arms, and along each radius of the urchin, form a powerful means of attachment.

On the most exposed shores, the battering of the waves prevents many seaweeds becoming established. Where there is some shelter, the

An example of a mutually beneficial partnership is illustrated here by a colony of sea squirts living on the shell of a shore crab. The detritus-eating sea squirts help to conceal the crab from predators by making it look like the rocks among which it lives; in return, they are carried along to new food sources when the crab moves to a different part of the shore.

large brown seaweeds can grow, attached to rocks by a special organ called a *holdfast*—an expanded disklike area at the base of the stem. Even when the main fronds have died or been eaten, the tough holdfasts are often found remaining stuck to pieces of rock or pebbles, and it takes considerable force to detach them. The shape of the fronds—long and ribbonlike or divided into narrowing branches—is well-adapted to withstand the tearing action of the waves.

The large colonies of mollusks and barnacles on rocky shores are a rich source of food for animals adapted to getting the soft parts out of their hard protective shells. The main invertebrate predator here is the dog whelk, a large marine snail that feeds on limpets, periwinkles, top shells, barnacles, and mussels. The dog whelk is a slow but persevering attacker, boring a narrow hole through the shell of its prey with its sawlike radula. It then inserts the radula, which is on the end of an extensible proboscis, and rasps out the meat inside the victim. The whelk also uses its radula to force apart the two shells of mussels. The color of a dog whelk's shell, which is determined by colored substances deposited in it, indicates the whelk's main source of food. Brownish purple is the color of a whelk that feeds chiefly on mussels; white, on barnacles. When a change of diet becomes necessary, the whelk changes its feeding method and the color of its shell also changes by the addition or removal of the blue-black pigment.

Some shore-living birds specialize in feeding on mollusks. The oyster catcher, conspicuous on a sandy shore but well concealed among rocks, with its boldly patterned black and white plumage, has a beak that is well-adapted for this purpose. It is bright orange-red, long and narrow, and flattened from side to side so that it can be used to chisel limpets off rocks and to pry open the tightly closed shells of mussels and other bivalves. Oyster catchers can be seen on all types of shore, and their beaks are equally useful for probing into sand or mud for worms and crustaceans.

The seaweeds of rocky shores provide food and shelter for a variety of quite large animals, including mollusks, crustaceans, starfish, and fish. They also create a miniature world of their own by providing an environment for some much smaller animals, many of which live in large colonies. One of the commonest brown seaweeds, the knotted wrack of the middle shore region, is often encrusted with pink tufts, which are colonies of the club-headed hydroid, a polyplike animal that belongs to the same group as sea anemones and coral. Serrated wrack is often covered by the white, netlike colonies of another group of simple animals, the sea mats or *Bryozoa*. And the dark purple seaweed carrageen, or Irish moss, frequently has large fleshy lobes of different sea mats growing on it, providing a habitat for many small crustaceans.

The tangleweeds that are exposed at low tide support a particularly rich encrusting fauna. Not only are their fronds covered by tufts of hydroids and films of sea mats, but they also form a base for sponges, sea squirts, tube-building worms, and barnacles. Among the most interesting of these animals are small segmented worms, which extract lime from sea water and enclose themselves in rock-hard tubes. The white spirals of these tubes, which look like handwriting scrawls, can be seen on weeds, rocks, and pebbles at low water and in rock pools.

Some sea squirts, like hydroids, are true colonial animals in that the whole company shares the food and oxygen taken in by each individual. They vary in color from rose-pink through red, purple, and blue, to yellow and orange; and the individual animals are often arranged in a regular starlike pattern in a gelatinous background substance. They are the food of sea slugs and cowries.

All these encrusting animals live permanently attached to the fronds and must go wherever the seaweeds are carried by the waves. It is obviously an advantage to them to live in large colonies where they are mutually interdependent and where the death or injury of a few does not harm the colony as a whole.

Apart from the rhythmic effects of the tides, conditions on a sandy shore are very different from those on a rocky coast. Instead of hard, immobile rocks, the animals and plants have the shifting sand as their background. Although sand is chemically the same as the rocks from which it has been derived through millions of years of grinding action by winds and waves, its finely divided structure gives it quite different physical properties. Organisms cannot attach themselves to the sand's surface and so they must burrow down into it. And in contrast to rocks, sand nearly always contains some moisture and food, even when exposed by the tide.

Perhaps the most striking feature of a typical sandy beach is the apparent lack of life on it. Very few seaweeds are found in this habitat because they have no roots with which to establish themselves, and a holdfast cannot grip tiny sand particles. A sandy shore actually supports a rich world of animals but most of them lie hidden in their burrows or among the debris of the high-tide line. For this reason vertical zonation is not as clearly marked as on a rocky shore.

Along the high-tide line pieces of weeds torn from rocks and washed up by the waves provide the main source of plant food and shelter. In and around the flotsam that is deposited at high water live a number of crustaceans known as beach fleas, which may be thought of as bridging the gap between sea and land because they can live either in or out of water. They spend the daytime in shallow burrows and hop down to the water's edge to feed on plankton at night. Ghost crabs, which blend so well with the sand that they seem to disappear when motionless, hunt for animal remains among the decaying weed, scuttling down into the sea to escape from their enemies.

The mid-tidal area of a sandy shore is the home of many burrowing mollusks—cockles, clams, and tellins—all of which use a muscular foot to dig themselves down into the sand. They are suspension- or detritus-feeders, using paired siphons to filter plankton and other tiny food particles from the water when covered by the tide. Burrowing crabs, also found in this area, hunt for small animals and debris in the sand.

On the lower shore, more completely marine animals living in burrows are exposed only at very low tide. Heart urchins, sea cucumbers, and bristle worms are detritus-eaters. Swimming crabs, starfish, sand eels, and flounders compete for plants and animals carried in by the tide.

Burrowing into sand is not difficult, but the problem for these animals is how to prevent the walls of their burrows from collapsing and thus

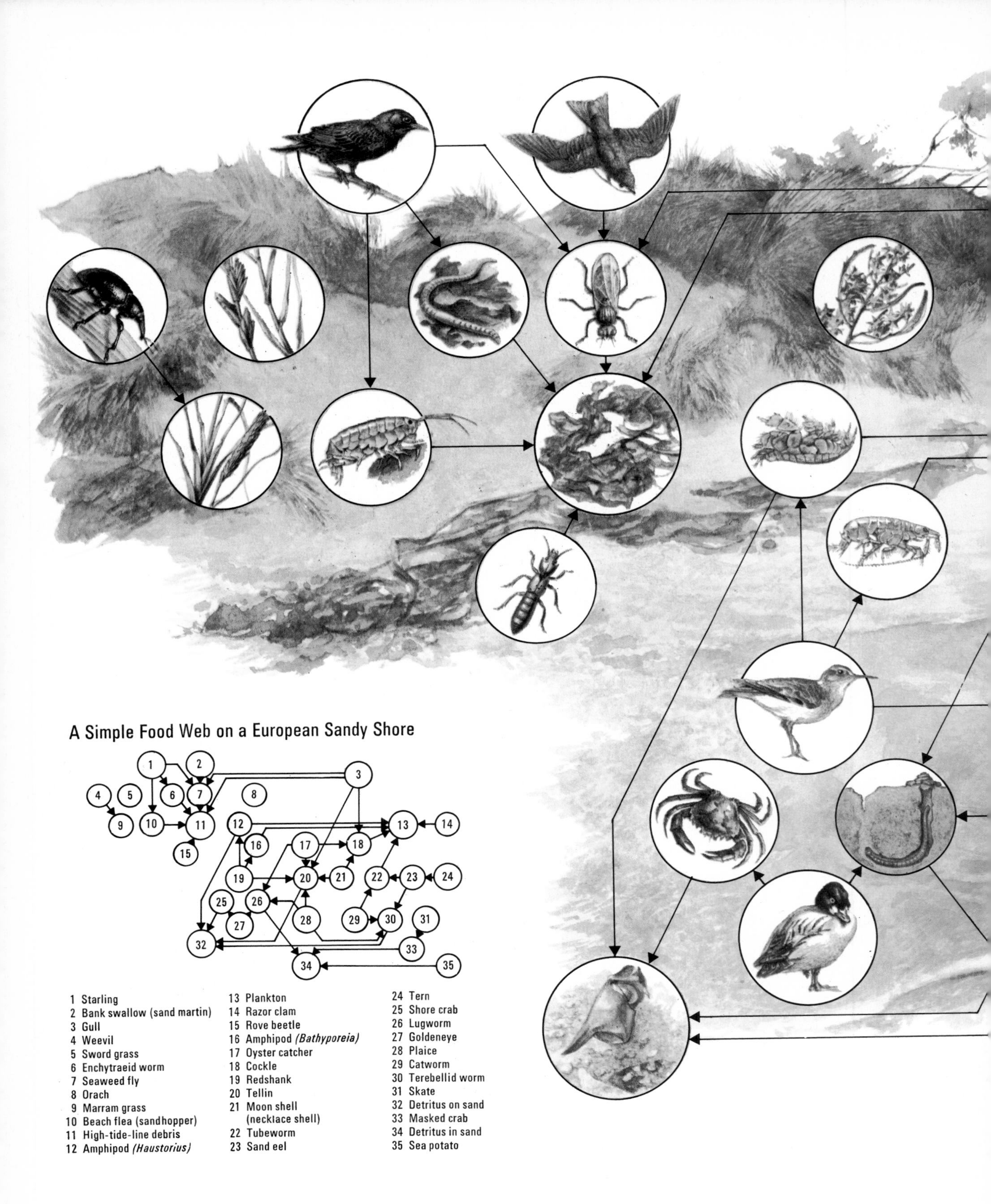

A Simple Food Web on a European Sandy Shore
1 Starling
2 Bank swallow (sand martin)
3 Gull
4 Weevil
5 Sword grass
6 Enchytraeid worm
7 Seaweed fly
8 Orach
9 Marram grass
10 Beach flea (sandhopper)
11 High-tide-line debris
12 Amphipod (Haustorius)
13 Plankton
14 Razor clam
15 Rove beetle
16 Amphipod (Bathyporeia)
17 Oyster catcher
18 Cockle
19 Redshank
20 Tellin
21 Moon shell (necklace shell)
22 Tubeworm
23 Sand eel
24 Tern
25 Shore crab
26 Lugworm
27 Goldeneye
28 Plaice
29 Catworm
30 Terebellid worm
31 Skate
32 Detritus on sand
33 Masked crab
34 Detritus in sand
35 Sea potato

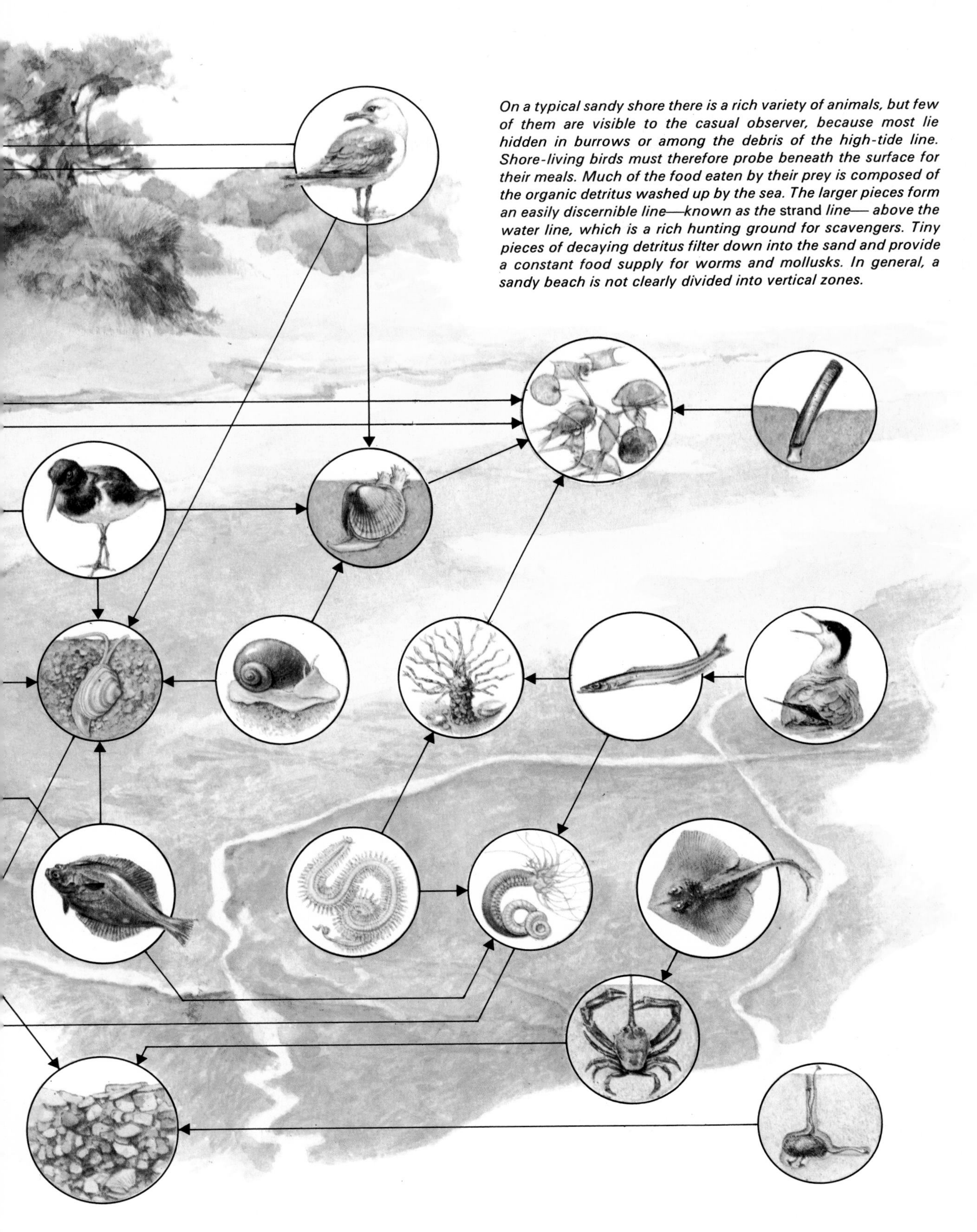

On a typical sandy shore there is a rich variety of animals, but few of them are visible to the casual observer, because most lie hidden in burrows or among the debris of the high-tide line. Shore-living birds must therefore probe beneath the surface for their meals. Much of the food eaten by their prey is composed of the organic detritus washed up by the sea. The larger pieces form an easily discernible line—known as the strand *line— above the water line, which is a rich hunting ground for scavengers. Tiny pieces of decaying detritus filter down into the sand and provide a constant food supply for worms and mollusks. In general, a sandy beach is not clearly divided into vertical zones.*

Characteristic crustaceans of sandy shores are the beach flea, or sandhopper (left), and the ghost crab (above), both of which are scavengers. To avoid its enemies, the sand-colored ghost crab scuttles sideways into the sea; the beach flea takes refuge by hopping and hiding among the flotsam.

cutting off their food and oxygen supply. This has been solved in various ways. Worms, for instance, produce in their skin a sticky mucus that cements the walls of their burrows, often so securely that they live in a complete tube of sand grains. Bivalve mollusks have long siphons that stick up just above the surface of the sand and carry incoming and outgoing water currents. The burrowing masked crab can convert its antennae into a breathing tube for use when it is completely buried in the sand.

A sandy shore is often occupied by large mixed flocks of wading birds. Their long legs and slim beaks allow them to probe the sand at low tide. There is a remarkable correlation between the length of the beaks of different kinds of bird and the depth of the burrows in which their prey live. The ringed plover has a beak about half an inch long, just able to reach the small marine snails that live barely under the surface of the sand. The one-inch beak of the sanderling is exactly the right length to dig up cockles, and is slightly flattened at the tip so that it can pry open their shells. The knot (a kind of sandpiper) feeds mainly on beach fleas, whose burrows, about one and a half inches deep, are within reach of its slightly longer bill. Redshanks can probe for small mollusks such as tellins, which burrow two inches down, and godwits can reach deeper bivalves at three to four inches. The curlew, with its five-inch down-curved beak, can get at the ragworms and lugworms whose burrows are too deep for the other birds to reach. Thus, each bird species is equipped to obtain one particular kind of animal food, and in this way many species of birds can live together in the same habitat.

Where the fresh water of a river meets the sea in a tidal estuary, a very special type of habitat is created. All the plants and animals that live in such an estuary must be able to withstand considerable changes in their environment frequently, as the incoming tide pushes salt water up into the mouth of the river, and then again as the water becomes less saline with the falling of the tide level. The adaptation of these organisms to changing salinities must therefore be a rapid one.

The main problem facing the plants of an estuary is the strong tidal flow. This prevents many plants establishing themselves, and so colonization of estuarine shores often takes the form of small hummocks around those that do succeed. As these hummocks grow they trap silt around them and form little islands around which the water must flow. The tidal channels between them are wide at first but gradually become narrower, restricting the flow of water. The increased force of the water then causes erosion of the islands. So from a network of wide, shallow ditches a typical salt marsh pattern of narrow, deep channels evolves under the influence of the estuarine plants. Some of the small islands join up, forming shallow salt pans, which fill at high tide and become exposed areas at low tide, accumulating a deposit of salt by evaporation.

One of the earliest colonizers in estuaries and salt marshes is the green alga *Enteromorpha*, which may form a thick carpet in brackish water. In the bottom mud live microscopic diatoms and one-celled algae in such huge numbers that they turn the surface layers green in the daytime, mainly at low tide, when they migrate upward toward the light.

Near the mouth of a typical European estuary the banks and shore may be covered by eelgrass. This is a grasslike flowering plant with fibrous roots, flat leaves, and tiny inconspicuous green flowers. Also characteristic of such estuaries and salt marshes is glasswort, which has fleshy cylindrical stems and leaves, and tiny white flowers. Above the level dominated by glasswort grows sea blite, and at the high-water mark tough cord grass forms dense tussocks. All of these plants effectively stabilize mud and sand with their widely branching fibrous root systems.

In the tidal channels, where there is always some water, live plants that spread down from the fresh water, such as shining pondweed and water milfoil, and others that spread up from the sea, such as bladder wrack, and tassel pondweed. All these can be regarded as brackish-water plants and survive in a wide range of salinities.

In most estuarine animals, the salt content of their cells can keep step with that of the water around them, and there are efficient regulating mechanisms to get rid of excess water entering at low tide, and excess salts at high tide. The shore crab and common periwinkle, which are often found in muddy estuaries, can survive in a wide range of salinities from almost fresh water to the

Above: the avocet, a bird of sandy and muddy estuaries, wades slowly forward on its long, stiltlike legs and sweeps its upcurved bill from side to side through the water in order to trap tiny animals, plants, and particles of detritus before eating them.

The common cockle is a bivalve mollusk, which uses its muscular pink foot to dig itself down into the sand. When buried, it gets its food by drawing a current of water through its shell and filtering out particles of detritus.

sea. Flounders, clams, and saltwater shrimps are less tolerant of low salinity and are not found as far away from the sea. Some animals, such as brackish-water prawns and the three-spined stickleback, live in widely varying salinities near the mouth of estuaries.

Changing salinity, then, is one key characteristic of an estuarine habitat. Another is the presence of large amounts of suspended particles of silt, carried down by the river and deposited on banks and shores. This is the origin of the wide mud flats that are a typical feature of many estuaries and salt marshes. Although nothing much is visible in mud flats the apparent lack of life in these fringes of the sea is deceptive: in reality life abounds here.

Many animals living in mud are filter-feeders, straining their diet from the water, or deposit-feeders, picking up food particles from the surface of the mud. These animals can burrow down into the mud for protection.

Estuarine animals include worms, bivalve mollusks, urchins and crabs burrowing in the mud, various insects and their larvae, as well as snails and crustaceans. The small conical depressions and intricately coiled worm casts visible on the surface indicate the entrances and exits of the subsurface burrows. The small pits show where various kinds of bivalves are hidden, and it is often necessary to dig to a depth of one foot or more to find such animals as soft-shelled clams. Because of their massive fused siphons, these creatures cannot escape predators by burrowing down fast—unlike the tellins and peppery furrow shells—and so they must remain deeply hidden.

Other depressions and worm casts indicate the presence of marine worms living in U-shaped burrows. These include lugworms, ragworms, peacock worms, and the sea mouse. The latter looks quite unlike any other worm, with its short, squat body covered by a mat of iridescent bristles. Like most bristle worms, the sea mouse takes in a mixture of mud, water, and decaying matter and leaves behind a worm cast of finely divided but indigestible mud particles.

Many wading birds can be seen probing the shallow water and mud of estuaries and brackish lagoons for food. Flamingos have the most highly specialized feeding technique of all the wading birds. Their boxlike bills have fringes of

In tidal mudflats, successful plants are those that can endure a range of salinity and changes in temperature and humidity caused by tidal waters. One such plant is glasswort (above), which establishes itself in the shifting mud and also stabilizes the mud by means of its widely branching fibrous root system.

bristles along the sides and the muscular tongue pumps water in and out. It is a strange sight to watch a flamingo, with its head inverted at the end of the snaky neck, busily sieving muddy water in which no food is visible. Greater flamingos feed on the deeper insect larvae, small mollusks and crustaceans, worms, and plant seeds, but the lesser flamingo, with its finer sieve, feeds exclusively on minute algae and diatoms.

The spoonbill uses a different technique. It sweeps its spatulate bill from side to side through the water in semicircles to catch small crustaceans, mollusks, and fish. The avocet holds the tip of its slender up-curved bill parallel to the surface, sweeping it from side to side as it walks slowly forward catching small invertebrates. Both of these birds feed only in shallow, muddy water. However, because this is usually the first kind of habitat to disappear when salt marshes and lagoons are drained for land reclamation, the spoonbill and avocet are now very restricted in their distribution.

In tropical latitudes, salt marshes and mud flats are replaced by mangrove swamps, which are home to a rich and varied collection of plants and animals. There are two types of mangrove swamp—the western mangroves on tropical Atlantic shores, and the eastern mangroves on the shores of the Indian and Pacific oceans. Although the actual plant and animal species found in the two types vary, the mangrove trees on which the swamps are built are closely related. The first colonizers are usually red mangroves, followed in succession by black mangroves and, behind them on the landward side, various palms and buttonwood trees. Each kind of tree has its characteristic community of smaller plants growing among its roots.

The two main obstacles to plant colonization in a muddy estuary—the tidal rise and fall of sea water, and the permanently waterlogged state of the mud—are overcome by mangroves in various ways. They overcome the first problem by producing seeds that sprout before they drop from the tree. The seedlings are therefore quite well developed when they fall, and so they can root and establish themselves quickly. The problem of waterlogging and consequent lack of oxygen for the roots is overcome by either aerial "stilt" roots hanging down from lower branches, as in the red mangrove, or knee roots, which grow above the surface of the mud and then down again into it, as in the black mangrove. A third type of root can be seen in Sonneratia species of mangrove, where conical breathing roots are sent into the air and produce a terrace of spear points. The ground layer of a mangrove swamp is therefore a tangle of these large roots with creepers, vines, and other plants clinging to them. The roots in a well-established swamp hold mud and create a stable base.

Mangrove roots provide a suitable environment for many kinds of animal, all adapted to conditions similar to those found in estuaries, of alternate flooding and desiccation and to varying salinities. One group of animals lives on or in the upper layers of mud, emerging from burrows at night or when the tide comes in. This group includes the fiddler crab (which gets its name from the enormous claw that the males have on one front limb), the little mudskipper (a fish that can "walk" and breathe in air), and the saltwater or crab-eating frog, the only amphibian known to live in salt water.

Adult saltwater frogs can tolerate a salinity of 29 parts per 1000, although they normally live in water of lower salinities, and the tadpoles are

A Simple Food Web in a Mangrove Swamp

1 Bark beetle
2 Bulbul
3 Firefly (adult)
4 Flying fox
5 Scale insect
6 Tailor ants
7 Sonneratia leaves
8 Woodpecker
9 Serpent eagle
10 Mosquito
11 Crab-eating macaque
12 Mangrove flower (Sonneratia)
13 Sunbird
14 Egret
15 Proboscis monkey
16 Heron
17 Darter
18 Mouse deer
19 Snail
20 Dead leaves
21 Crab-eating frog
22 Mud crab
23 Winkle
24 Algae on leaves
25 Land crab
26 Debris
27 Swimming crab
28 Crab-eating crab
29 Water snake
30 Firefly (larva)
31 Mudskipper
32 Catfish
33 Marine snail
34 Detritus and microscopic animals
35 Oyster
36 Fiddler crab
37 Mud prawn
38 Barnacles
39 Mixed plankton
40 Clam

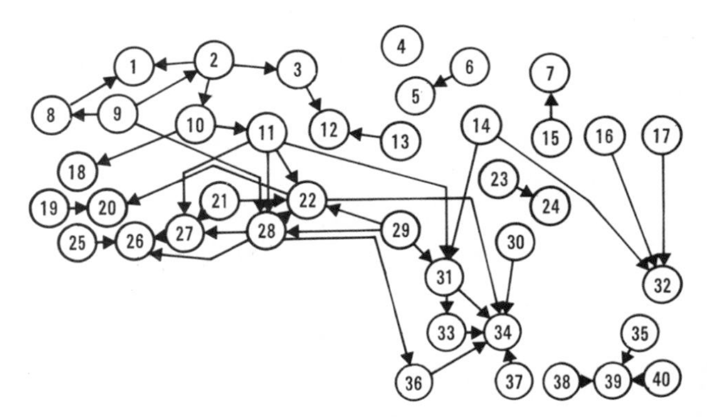

The food web of a tropical mangrove swamp is dependent on the aerial roots that enable the mangrove trees to grow in water-logged soil and also help to stabilize the mud. Animals living on the roots and in the mud are eaten by birds, reptiles, and mammals (as shown by the arrows in the key), many of which visit the branches in search of fruit or to catch fish at high tide.

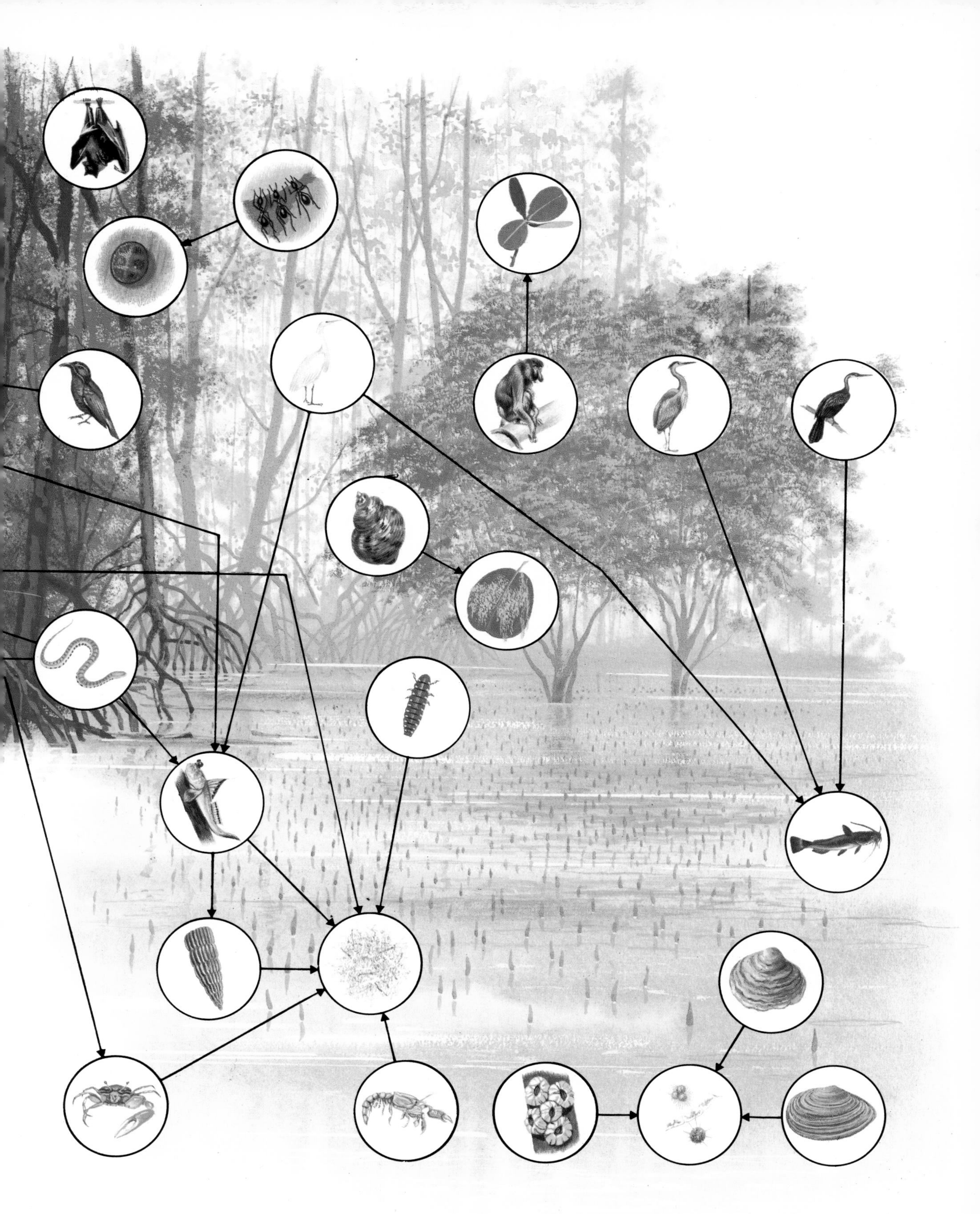

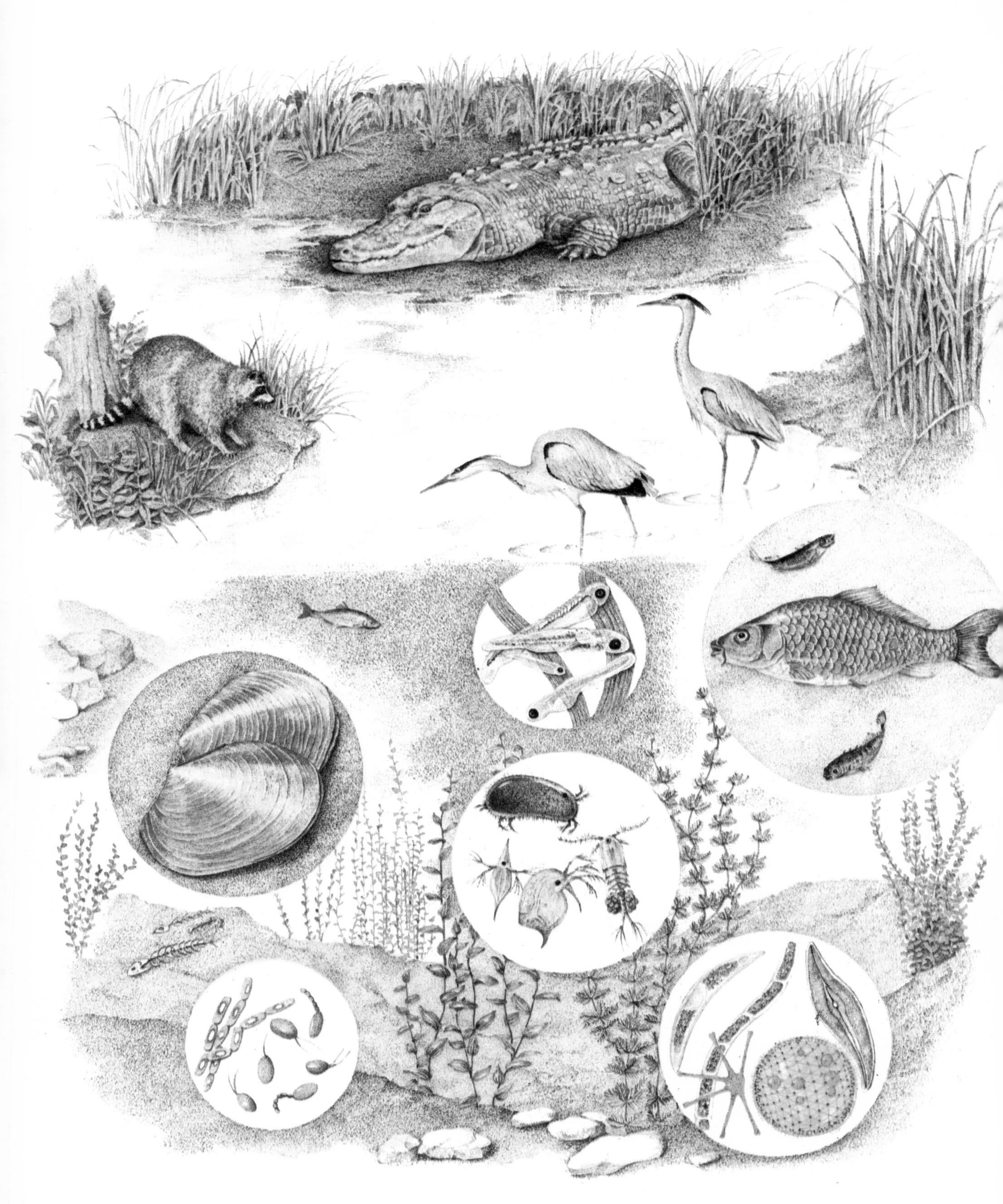

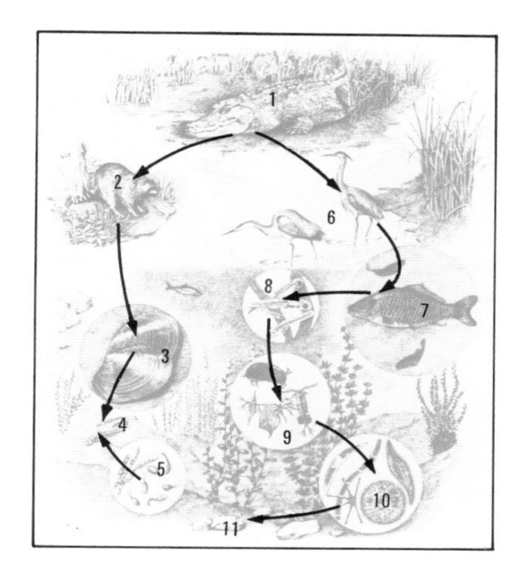

Alligator Food Chains in a Salt Marsh in Florida

1 Alligator
2 Raccoon
3 Mussel
4 Detritus
5 Bacteria and fungi
6 Heron
7 Fish
8 Fish larvae
9 Crustaceans
10 Plant plankton
11 Mineral salts

Where a river flows into the sea in a subtropical region, characteristic food chains can reach from the smallest microorganisms up to mighty creatures such as the alligator. Two such food chains in a Florida salt marsh are shown here. Further links can be drawn between these two chains and they interlink with many other chains to create a complicated food web. Bacteria and fungi continuously break down the decaying particles in the bottom silt. This produces nutrients for the planktonic plants that tiny crustaceans, larval fish, and mussels feed on. Adult fish eat the smaller animals, and herons and racoons are expert at catching fish. The alligator completes both chains.

able to live in a solution even more concentrated than sea water. The frog does not drink water, but its skin allows water to pass through, so it is just as vulnerable to water loss by evaporation as other amphibians. As the water becomes increasingly saline, the frog's skin is able to absorb salt actively from the surrounding water in order to bring about an increase in blood plasma concentration to keep pace with the external salinity. This minimizes the gain or loss of water through the skin.

Various animals live attached to the mangrove roots, including worms, barnacles, mussels, oysters, and sea squirts; and small fish such as blennies live in crevices among the roots. Some of these creatures are *euryhaline*—that is, they can tolerate a whole range of salinities. Others are more particular and are found only in zones that provide just the right conditions.

Some of the animals found in mangrove swamps are terrestrial in origin. They either live permanently among the branches of the trees or visit the roots and mud at low tide. There are, for instance, the leaf-eating insects that feed in the trees. They provide food for a variety of lizards, which in turn are eaten by tree snakes. At night parrots and other birds roost in branches, having fed in nearby fruit-bearing trees during the day. When the tide is out, saltwater crocodiles lie on the mud, waiting for unwary birds and small mammals to come within reach of their jaws. Wild pigs and monkeys visit the mangroves in search of delicacies such as frogs, crabs, and other shellfish. Deer browse on the plants growing among the arched roots. Ibises, egrets, rails, and many other birds probe the mud for mollusks, and spoonbills feed in the shallow water.

Some marine animals visit the mangroves when the tide rises. Gobies, mullet, and other fish graze on algae growing on the prop roots that arch outward from the main stem and plunge into the mud. This increased fish population attracts birds such as cormorants, boobies, and kingfishers. The aerial roots provide excellent perches from which these fishing birds can dive after their prey.

Among the strangest phenomena in the animal kingdom are the great migrations between fresh water and sea of certain fish, the best-known being the salmon and the eel. This requires physiological adaptations to enable the animals to cope with the differing salinities of the water.

Through the evolutionary process the body fluids—mainly blood plasma and cell protoplasm—of aquatic animals have become adapted to the media in which the creatures live. In most marine invertebrates the salt content of the cells and blood plasma is close to that of the salt water in which they live. For vertebrates, however, the situation is more complicated. Sharks and rays have body fluids that are slightly more concentrated than sea water and they must therefore excrete by way of their kidneys the extra water that passes in through the thin-walled gills. Most bony sea fish have body fluids less concentrated than sea water. These fish drink sea water, and therefore, to maintain the correct balance of salts and water within, they must get rid of unwanted salts by excreting them back into the sea from their gills.

The salmon, a bony fish, starts its life in fast-flowing rivers, where the spawn is laid in small depressions, or *redds*, which are made by the female fish lashing her tail in the gravel of the river bed. After the young fish hatch they develop until they reach a length of about four inches, when they are called *parr*. Like salmon eggs, parr can survive only in fresh water.

After one year for Pacific salmon, the parr becomes *smolt*. By the time the smolt stage is reached, the young fish has acquired salt-secreting cells in its gills and other special body functions to help it cope with the salt in the sea water. It is at this point that the long journeyings begin, prompted by a hormone produced in the pituitary gland. The smolts swim down river and into the sea, where they spend a year as *grilse*, their next stage, feeding on the small shrimps of the plankton. At the end of this period Pacific salmon are mature adults and they then make the long return journey to their birthplace to spawn. The fish battle their way up waterfalls and against the current, guided by a strong homing instinct. Until quite recently, the fact that salmon that had been given an identifying tag as parr were always found as adults back in their native rivers was a mystery. Now we know that they have extremely sensitive noses and are able to recognize the water from their home rivers by smell. Adult Pacific salmon use up so much energy and become so emaciated and battered on the journey upstream that they all die after releasing their eggs and sperms.

Atlantic salmon have longer lives than their Pacific relatives, and they spend two or three years as parr before they become smolt and swim to the sea. The majority of adult Atlantic salmon die after spawning but the survivors may return twice or even three times in successive years to their home rivers to spawn again. All adult salmon make the journey from the sea up the rivers so quickly that they must stop to rest, and allow their sensitive kidneys to recover from the extra activity, in the first large calm pools they reach. It is in these salmon pools that they often become easy prey for fishermen.

Another bony fish, the eel, migrates in the opposite direction from the salmon. The eggs hatch in the sea and young eels, like other bony fish, drink sea water to maintain their salt and water balance. For many years naturalists were puzzled because young eels were never found in rivers where adults were common fish. This has led in times past to the belief that eels were born spontaneously from mud or from horse hairs dropped

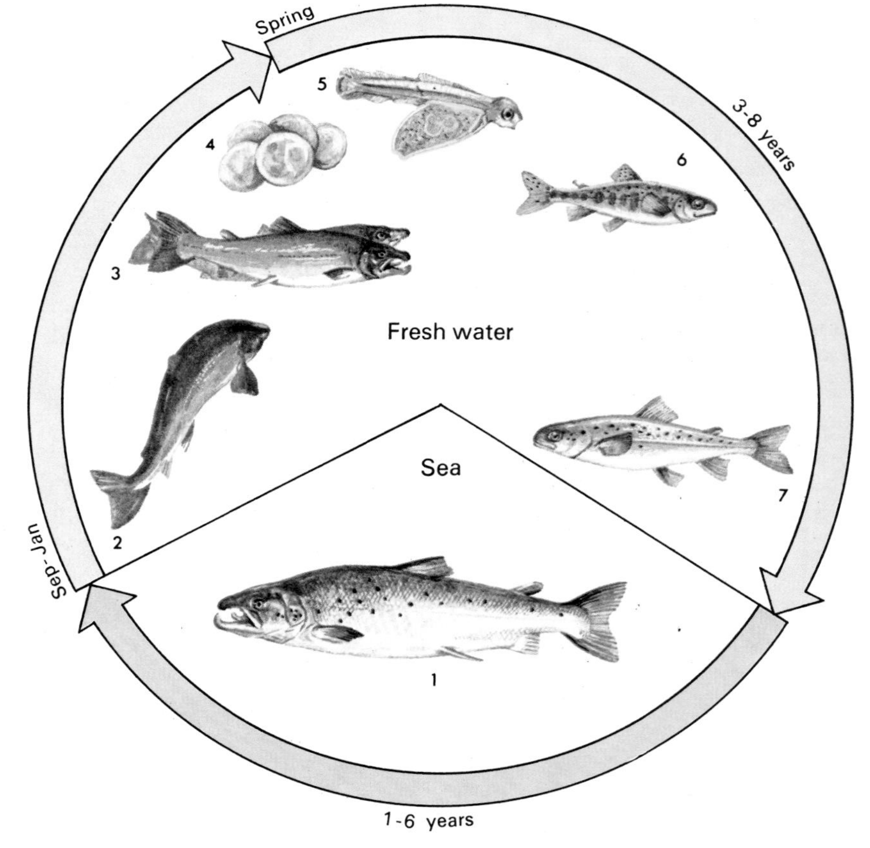

Right: the life cycle of the Atlantic salmon. Mature salmon (1) migrate from the sea to the rivers of their birth, leaping rapids on the way (2). They arrive and spawn during the winter months. The pink males (3) fertilize the eggs (4) and in the spring the fry (5) hatch. The next 3 to 8 years are spent in fresh water, during which time the salmon pass through the parr stage (6) shown above. They then become smolt and the young salmon are ready to travel to the sea. Between 1 and 6 years are spent in the sea before the mature salmon return to fresh water to spawn.

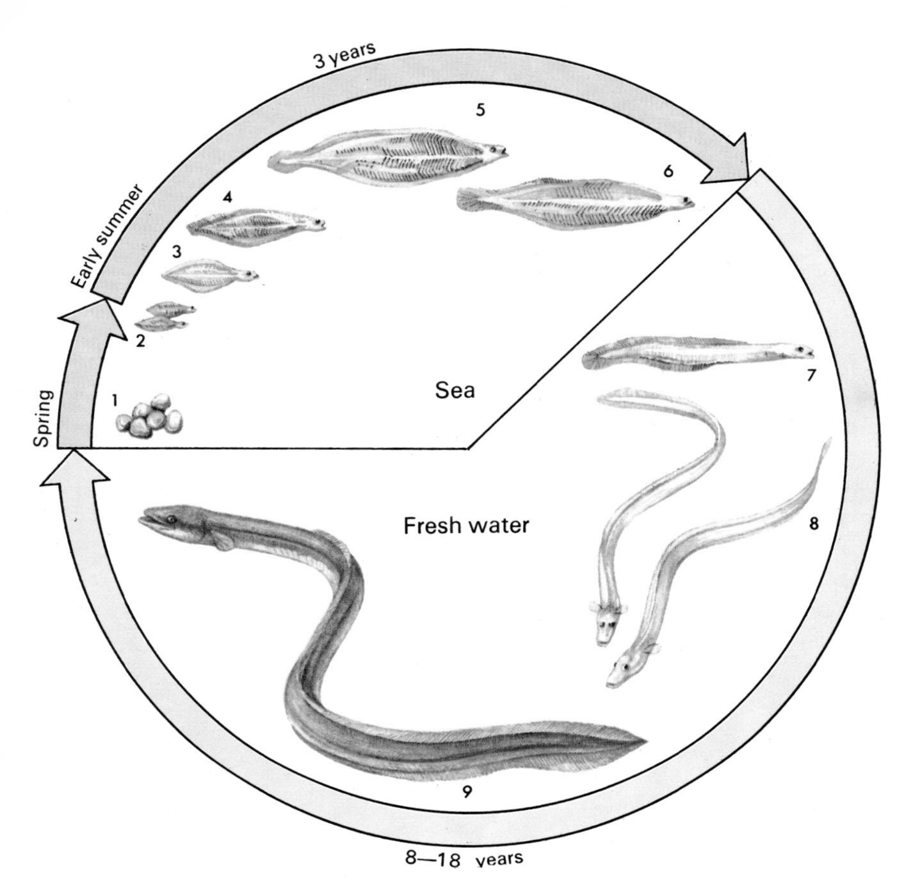

Left: the life cycle of the European common eel. Adult eels spawn in spring in the Sargasso Sea (1), and then die. The tiny larvae, or leptocephali, *hatch in early summer (2). During the next three years, while they grow and gradually change shape (3-6), they drift in the Gulf Stream to European coastal waters. Here they change into transparent elvers (7) and by the time they reach their destinations of rivers and streams they have grown considerably (8). The adult eels (9) remain in fresh water for many years, the males staying for about 8-10 years and the females for 10-18 years, before returning to the sea to spawn.*

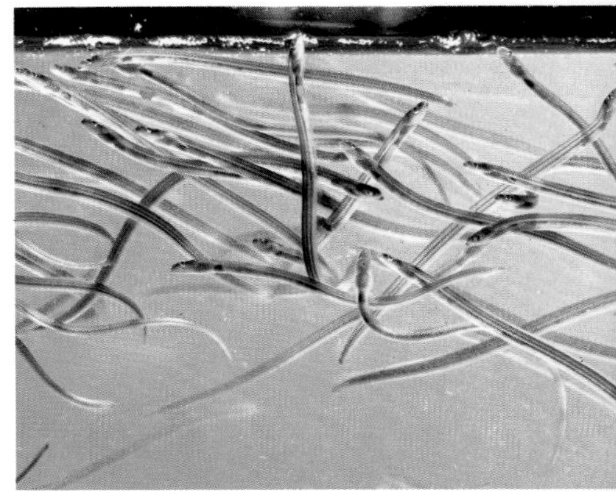

into the water. Equally mysterious was the little animal called *Leptocephalus*—a small, transparent, leaf-shaped fish found in many parts of the Atlantic. It had never been kept successfully in an aquarium or been observed to breed. Gradually the twin mysteries were solved, beginning with the observation of a few *Leptocephali* as they changed into elvers (young eels).

We now know that adult eels spawn in quite a small area of the Atlantic Ocean, the Sargasso Sea. From here the eel larvae—which is what *Leptocephali* really are—travel over 2000 miles in the North Atlantic Drift and the Gulf Stream, to the coasts of Europe and North America. This journey, prompted by the pituitary hormone, takes at least three years, during which time the young eels grow from about a quarter of an inch to about two inches long. When they reach the mouths of European and North American rivers they change rapidly into long, narrow, black elvers and spend the next 8 to 18 years as freshwater fish. The eel's journey from the sea to fresh water is a slow one compared with that of the salmon returning to spawn. Over a period of months its body can make a gradual change from getting rid of extra salt to getting rid of excess water. To help this process, when they are in fresh water eels do not drink and their skin produces abundant waterproof mucus. Adult eels up to three feet in length return to the Sargasso Sea to spawn at the end of their lives, and the whole cycle begins again. There remain two mysteries. First, why is it that very few mature eels are ever caught in the sea? And secondly, how do they navigate from rivers as far apart as Iceland and North Africa back to the same small area of the ocean where their lives began? Apart from the long annual migrations undertaken by a few species of birds, the distances traveled by an eel in its lifetime are certainly the greatest any animal covers.

Shores, lagoons, and estuaries—the fringes of the sea—are some of the most intensively studied habitats in the world. This is partly because they are generally accessible and easy to study, but the main reason is the great variety of organisms that live in them. These organisms display a range of special adaptations to very difficult living conditions at the meeting places of salt water, fresh water, and dry land.

The Balance of Life

This book has attempted to paint a broad picture of the oceans and the life within them. We have seen how each living organism is adapted for survival in its own environmental niche. Life in the oceans is of particular interest to man because the seas have been the cradle of evolution. Perhaps 90 per cent of all the animals and plants that have ever existed have lived there. In one way or another, all have depended on and contributed to the total oceanic environment.

For countless millions of years, the seas that cover 70 per cent of the earth's surface have acted as a vast reservoir, making life on the land possible. Water drawn up by evaporation from the sea condenses in clouds, is carried by winds, and descends as rain or snow on land areas. As a result of the heat-holding capacity of the oceans, many places that would otherwise be uninhabitable have a reasonable climate. As well as being the universal source of water, the oceans are also the universal filtering and purifying system; they receive filthy water from rivers and cities and return it as pure rain. Animals and plants and their remains are continually being eaten, broken down, and decomposed, but the sea remains fit to live in as long as a balance of producers and consumers is maintained.

For man the sea provides a varied larder of protein-rich food, although at present he does not utilize more than one fifth of its potential supply. Indeed man is far from appreciating its value to him. Out of laziness and greed he constantly does great harm to the sea and its inhabitants. He snatches from it the most easily obtained and economically valuable products and in the process he kills off the great whales, walruses, and turtles; overfishes the continental breeding grounds of herring, cod, and plaice; and breaks up the seabed in his hunt for oil.

In return for taking all these riches from the oceans, what has he put back? Ever since men began to live in towns and cities, the sea has been their sink, refuse-disposal unit, and sewer. In recent years in particular, ever-increasing quantities of untreated sewage, poisonous chemicals, radioactive waste, artificial fertilizers, and plastic garbage have been pouring into the sea.

Until very recently, no thought was given to the

There is nothing basically wrong with industrial fishing even though it takes vast quantities of food out of the world's oceans—often by means of trawl nets like the one being hauled aboard at left—because the seas are capable of replenishing themselves. But modern man has dangerously tipped the balance by overexploiting nature's resources: overfishing breeding grounds, killing off the great whales (above), drilling the seabed for oil, and using the sea as a sewer.

possible results of our self-centered treatment of the oceans; "out of sight, out of mind" has been the general attitude. Then things began to go wrong. Oil from wrecked tankers was washed up onto favorite summertime beaches—an ugly, thick, black mass, strewn with dead fish and seabirds. Mussels that had accumulated cholera bacteria were found to be the cause of a serious cholera outbreak in holiday resorts on the Italian Adriatic, and were soon declared unfit to eat in many other places where they had once been safe and plentiful. Blue whales became so scarce that it was uneconomical to go on hunting them. Fishing in estuaries and bays virtually ceased as great numbers of fish died in the polluted water. As an elderly Venetian remarked during the 1973 cholera epidemic, "We have been pouring filth into the sea for years. Now the sea is returning some of it to us." Only when the shock of such happenings causes a public outcry do our governments and their advisers begin to realize that the condition of the oceans can no longer be ignored, and that our treatment of them must be planned and controlled in the future.

A paradox of the sea is that as we learn more about it, the danger of misusing and exploiting it becomes greater. Left to itself, for instance, a coral reef is largely self-regulating for the many animals living in and around it. Once people discover how interesting and beautiful a reef is, they begin to carry away the coral and other trophies. Shell collectors remove hundreds of predatory mollusks, some of which prey on the coral-eating crown-of-thorns starfish. Underwater fishing enthusiasts kill large numbers of puffers and triggerfish, which also feed on the crown-of-thorns. By removing these checks on its numbers, man allows the starfish to multiply rapidly at the expense of the coral. In a short time, the reef becomes a ruin—a mass of dead

Although the sea is able to break down all kinds of organic matter, unlimited quantities of sewage or garbage could make it unfit to live in. Far from being a bottomless waste-disposal unit, the water periodically cleanses itself by redepositing refuse on land, as shown in these pictures of polluted harbors.

49

Ironically, an appreciation of the beautiful creatures in the sea can greatly harm the environment. Do you covet the pretty shells at left? You may be helping to kill off the mollusks that prey on the crown-of-thorns starfish (above), which destroys acres of coral every year by feeding on the living coral polyps.

coral heads, unable to replace themselves, and unable to support the few remaining fish.

There is also the danger of taking for granted the apparently endless supply of popular food fish. Most such fish breed in the continental shelf regions, and because it is these waters that receive the main outpourings of pollutants, the food fish population is quickly and drastically affected by man's thoughtlessness. There are already signs that in the main fishing grounds of the North Atlantic the catch of some kinds of fish is falling. With the great numbers and variety of animals involved, the complex food web in these areas is very easily upset by the modern practice of fishing with fine-mesh nets to catch young fish. These are then turned into animal feeding stuffs instead of being permitted to live and reproduce their kind. Paradoxically again, the coastal waters on which we most depend for foodstuffs are the most vulnerable to our depredations.

Although the mid-ocean areas are farther from the direct effects of man-made disturbance, we should never forget that in fact the seas and oceans of the world form a continuous and unified system. If we upset the relationships in one part, we are assuredly endangering the whole. And whatever harms the life of the oceans will inevitably harm us, too. This is why it is well for all of us to know as much as we can about life in the sea. The more we know about it, the more we respect and value it.

Index

Page numbers in *italics* refer to illustrations or captions to illustrations.

Picture Credits

Key to position of picture on page: (B) bottom, (C) center, (L) left, (R) right, (T) top; hence (BR) bottom right, (CL) center left etc.

Cover: K. Gunnar/Bruce Coleman Inc.
Title page: J. J. Meusy/Pitch
Contents: Jen & Des Bartlett/Bruce Coleman Inc.
9 John Dominis, *Life* © Time Inc. 1975
10 Dr. D. James/ZEFA
14 Paul-Emile Victor/Pitch
15 Bryan Alexander/Aspect
16 Heather Angel
17 National Audubon Society
18 George Holton/Photo Researchers Inc.
19(TR) Heather Angel
19(B) Neville Fox-Davies/Bruce Coleman Ltd.
20 Peter David/Photo Researchers Inc.
21 © Douglas P. Wilson
22 Colin Doeg/Seaphot
23(T) Allan Power/Bruce Coleman Inc.
23(B) © Douglas P. Wilson
24(T) G. Vienne/Pitch
24(B) Albert Visage/Jacana
25(L) David & Katie Urry/Bruce Coleman Ltd.
25(R) Jacana/Boisson
26 after N. J. Berrill, *The Life of the Ocean*
28 Ron Taylor/Ardea, London
30(L) Peter Hill, A.R.P.S.
30(R) © Douglas P. Wilson
31(R) Russ Kinne/Photo Researchers Inc.
34 Heather Angel
35(T)(BR) David/Seaphot
37(TL) Chaumeton/Jacana
37(TR) Allan Power/Bruce Coleman Ltd.
37(B) P. Morris Photographics
38(T) Stephen Dalton/NHPA
38(B) Noailles/Jacana
39 © Douglas P. Wilson
40 Grossa/Jacana
41 Russ Kinne/Photo Researchers Inc.
42(L) © Douglas P. Wilson
43 Ron Church/Photo Researchers Inc.
44 Jeff Meyer/Animals Animals ©1974
47 Grossa/Jacana
48–9 © Douglas P. Wilson
50 P. David/Seaphot
51 Chaumeton/Jacana
52(T), 53 Heather Angel
52(B) Peter David/Seaphot
55 Peter Hill, A.R.P.S.
56 Peter Scoones/Seaphot
58(L) Charlie Ott/Bruce Coleman Ltd.
59 C. Ray/Photo Researchers Inc.
60 Peter Hill, A.R.P.S.
61(R) Heather Angel
62 Popperfoto
63(L) Peter David/Photo Researchers Inc.
63(R) Peter David/Seaphot
65 Chaumeton/Jacana
67 Tom Myers/Photo Researchers Inc.
68 Tom McHugh/Photo Researchers Inc.
69(R) A. Thau/ZEFA
71 James Tallon/NHPA
72 Tom McHugh/Photo Researchers Inc.
73(T) © Douglas P. Wilson
73(B) Geoff Harwood/Seaphot
74 Tom McHugh/Photo Researchers Inc.
75 Jen & Des Bartlett/Bruce Coleman Inc.
76 FPG Inc.
78–9 Black Star, New York
80 Peter David/Seaphot
81(R) Jane Burton/Bruce Coleman Ltd.
82–3 Heather Angel
84 © Douglas P. Wilson
86(B) Jane Burton/Bruce Coleman Ltd.
87 Eliott/Jacana
88 Mario Fantin/Photo Researchers Inc.
89(TR) de Klemm/Jacana
89(BR) C. Ray/Photo Researchers Inc.
91 Mario Fantin/Photo Researchers Inc.
92 George Leavens/Photo Researchers Inc.
94–5(T) Francisco Erize/Bruce Coleman Ltd.
94(B) Trans-Antarctic Expedition
97 Leonard Lee Rue/Bruce Coleman Ltd.
98 Sven Gillsäter/Bruce Coleman Ltd.
99(R) J. Prevost/Jacana
100 Jane Burton/Bruce Coleman Ltd.
101 Allan Power/Bruce Coleman Ltd.
102 Heather Angel
103(R) Peter Vane/Seaphot
105 Russ Kinne/Photo Researchers Inc.
106(T) M.P. Harris/Bruce Coleman Ltd.
106(B) Allan Power/Bruce Coleman Ltd.
107(BL) H.A.E. Lucas/Spectrum Colour Library
107(R) Popperfoto
108 Peter David/Seaphot
111 J. Barlee/Frank W. Lane
112(L) Peter Hill, A.R.P.S.
113(T) Vasserot/Jacana
113(B) Charles R. Belinky/Photo Researchers Inc.
116 Heather Angel
117 Jane Burton/Bruce Coleman Ltd.
118 Anthony & Elizabeth Bomford
120 Heather Angel
124(L) Jane Burton/Bruce Coleman Ltd.
124(R), 125(R) Heather Angel
125(R) M. Weichmann/Frank W. Lane
126–7 Heather Angel
132(L) P. Morris Photographics
132(R), 133(L) Cathy Jarman, *Atlas of Animal Migration,* Aldus Books Limited, London, 1972
133(R) Jane Burton/Bruce Coleman Ltd.
134 Anthony & Elizabeth Bomford
135(R) R. J. Griffith
136 © Carl. E. Ostman AB
137 United Press International (U.K.) Ltd.
138 P. Morris Photographics
139(R) G. R. Roberts, Nelson, New Zealand

Artist Credits

David Nockels 12–3, 32, 128–9; Tony Swift 114–5; Joyce Tuhill 26–7, 130–1; Peter Warner 122–3.

We would like to thank Dr. Michael Hassell and Dr. Stuart McNeill of Imperial College, London, for providing the original sketches for the food web diagrams on Pages 32, 114–5, 122–3, and 128–9.

THE LIVING WATERS

Part 2
Rivers and Lakes

by Peter Credland

Series Coordinator	Geoffrey Rogers
Art Director	Frank Fry
Design Consultant	Guenther Radtke
Editorial Consultants	Donald Berwick
	David Lambert
	Malcolm Ward
Series Consultant	Malcolm Ross-Macdonald
Editor	Bridget Gibbs
Copy Editors	Maureen Cartwright
	Joanna Jellinek
	Damian Grint
Research	Naomi Narod
	Julia Hutt
Art Assistant	Vivienne Field

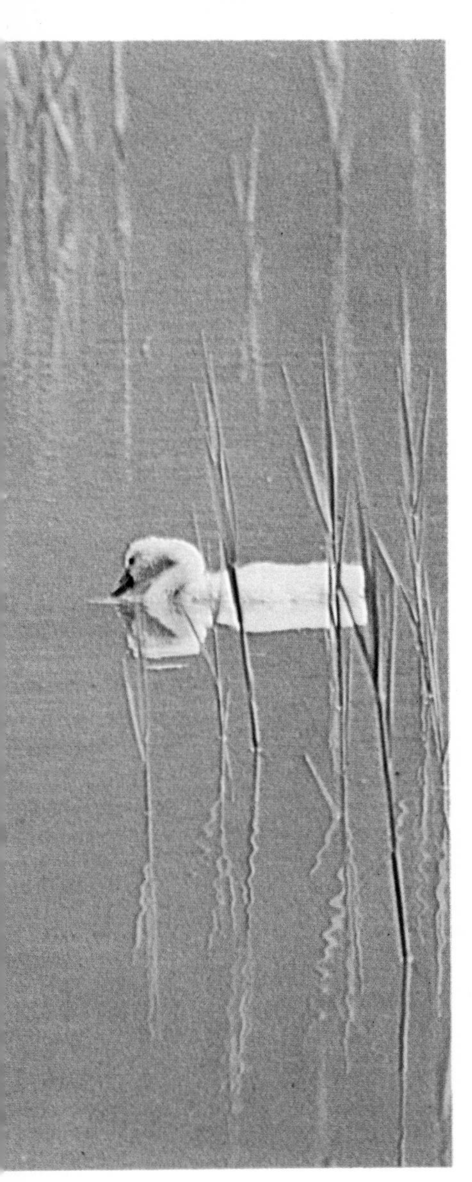

Contents: Part 2

Editorial Advisers

Introduction

Our associations with the rivers and lakes around us fall broadly into two categories, which may be headed "pleasure" and "necessity." Under the first we can include such activities as fishing, boating, and water-skiing, or, of course, simply enjoying the beauty and tranquillity of water. Under the heading "necessity" we include the water that is used for drinking, for generation of power, for a large number of industrial processes, and for the transportation of domestic and industrial waste to the sea.

Man's insatiable appetite for fresh water to satisfy all his needs has focused his attention on the interrelationship of these two categories and on the consequences of his abuse of one of his most precious resources.

In this book we take a look at this vitally important freshwater resource. One of the first things we discover about our rivers and lakes is that they are living communities of distinctive animals and plants and we shall be looking at some of the more typical of these communities and their habitats. Our survey of freshwater communities will reveal many of the significant ways in which plants and animals have adapted to a "competitive co-existence." It is a balance that is essential to the well-being of a freshwater community. The threat of new pressures created by the ever greater industrial, domestic, and recreational demands for fresh water in the remaining years of this century may have widespread effects on our rivers and lakes and it will require all our vigilance to protect and preserve them. We can start now by examining more closely how an undisturbed community survives. The lessons we learn should help us not only to protect the freshwater environment for its own sake, but also help us to participate more fully in the community life of our planet.

The Freshwater Environment

Fresh water has a unique fascination for mankind. The magnificence of a waterfall cascading hundreds of feet into a chasm, the gurgling of an upland brook, the sleepy meandering of a river, or the tranquillity of a lake—these are images to inspire the imagination of us all. How many of us when we go out for a walk can resist the chance to follow a river, even if only for a short distance; and who is not attracted by the sound of running water when climbing in the hills? Why do so many people keep fish in a tank, or spend money on a garden pool or fountain?

The desire to dabble at the water's edge is a feature of almost every child's life. Throughout the summer, young boys and girls equipped with nets and jars flock to lakes, rivers, or canals. The hunt is for any small animal, but it is usually the telltale flash of silver, indicating the presence of a minnow or stickleback, that gives rise to the greatest excitement. The dipped net often catches only a mass of green blanketlike weed, some water snails, and a few insects that look like mobile suits of armor in miniature. These little animals are often overlooked, because most people do not realize that they are more fascinating to observe than many a fish. Perhaps it is because they move so slowly that they are neglected.

Having passed the age at which fishing for minnows is a preoccupation, a great many people take up angling. In many respects less efficient than the juvenile net, the line always offers the hope of larger prey. However, one wonders how many anglers invest in expensive equipment in order to catch fish and how many in order to leave their everyday life and sit peacefully at the water's edge, watching the plants and animals in and around the lake or river. It is perhaps a reflection on our "civilization" that many anglers hook the artifacts of our way of life rather than the prey they hope for. How many have not caught tin cans, plastic bags, or pieces

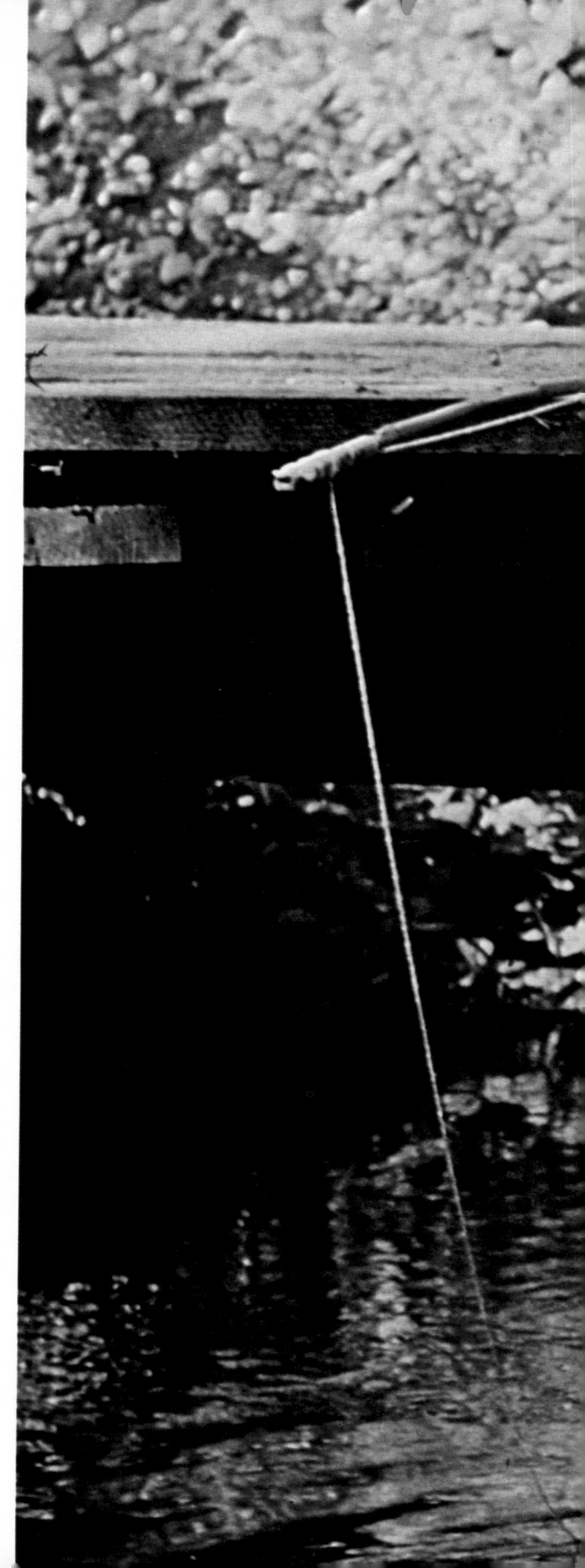

To the boy on the bridge the stream below is full of mystery. What unseen creatures lurk in its depths? How do they live? The boy's rod is less a hunter's weapon than a scientist's probe. With its aid he may learn at least some of the secrets to be found in the fascinating world of fresh water.

Food Pyramid for Silver Springs, Florida

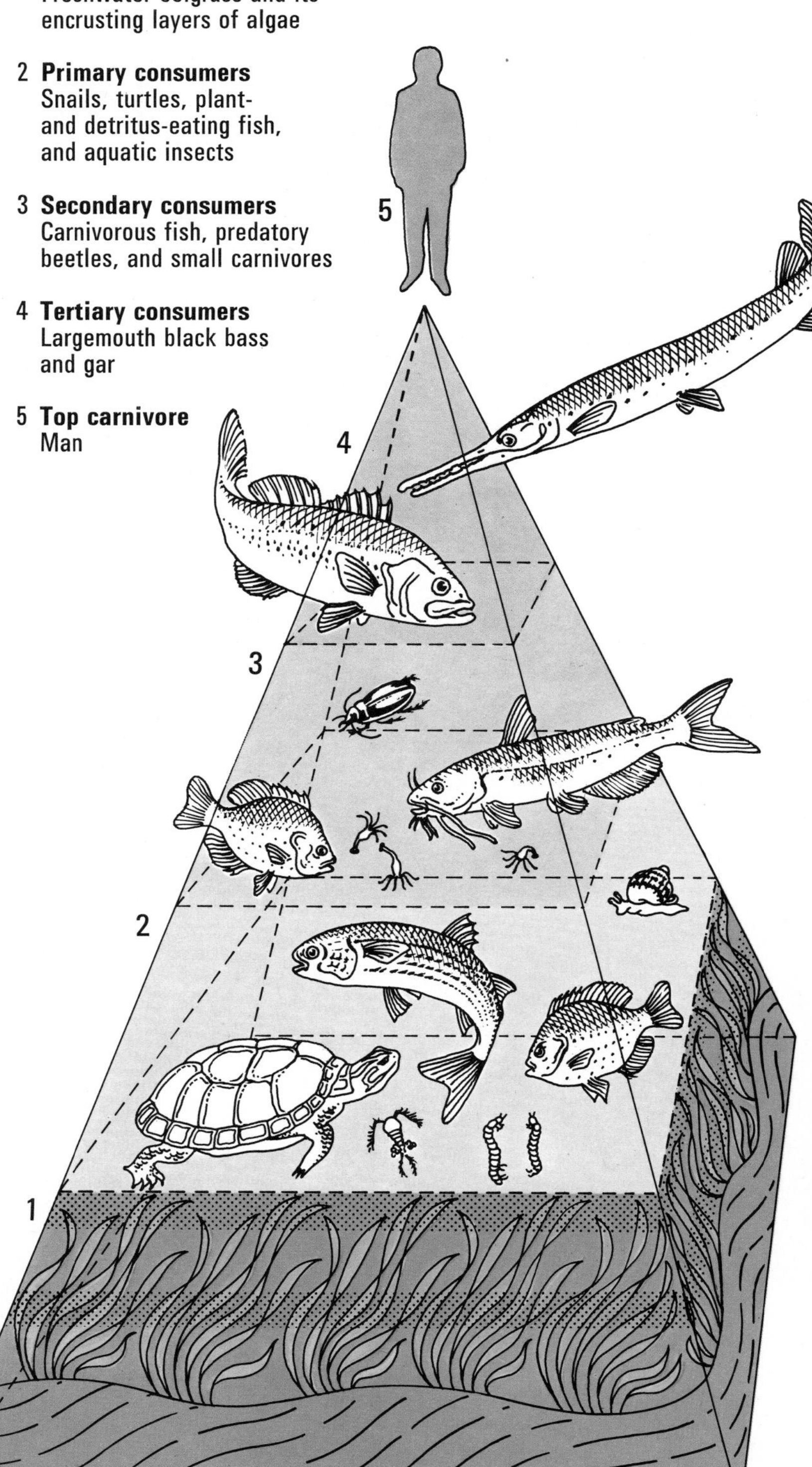

Left: a typical freshwater food pyramid reminds us that plants (1) form the broad base upon which all animal life depends. Next in abundance are the plant-eaters (2). Some become food for a smaller meat-eating group (3). These meat-eaters are eaten by larger but less numerous carnivores (4), in turn consumed by a top carnivore (5). Each creature can convert only a small percentage of the food it eats into the substance of its own body. This is why the food pyramid narrows as it rises.

Above: fish-farm workers collect trout eggs, then raise the fry in artificially disease-free, food-rich conditions to produce dense fish-farm shoals (right). In nature, usually not more than one egg in thousands survives to become an adult trout, and one trout may need the food resources of a large area.

of waste paper, or even the "old boot" of children's books? It is a tribute to the versatility of animals and plants in the water that even these items are frequently found to harbor innumerable living organisms.

From earliest childhood we are aware that water as a medium is very different from the air in which we live. To the casual observer the fish may appear to be the lords of this world beneath the water. But they are only a small part of a complex web of life composed of a wide range of creatures, a few of them very similar to those with which we are familiar in the garden, field, or hedgerow—although most of them are quite different. The history and life style of each of these animals intertwines with those of many others. Animals depend on each other for a variety of reasons, one of the most obvious being the need for food. As the lion hunts zebra or antelope, so the fish feeds on insects or worms. The situations are almost parallel. In nearly every case the prey is smaller and more numerous than the predator. Thus, for every single fish there are dozens—perhaps hundreds or even thousands—of insects and worms on which the fish can feed. This theme of larger consuming smaller is repeated throughout the community down to the level of tiny creatures living in the soil of the plains or the muddy bottom of a lake. There are a few exceptions, but not enough to upset the general rule. Mostly they involve small animals acting in concert. For example, wild hunting dogs may work in packs to bring down a zebra, and a large fish or bird may be torn apart by a shoal of piranha fish.

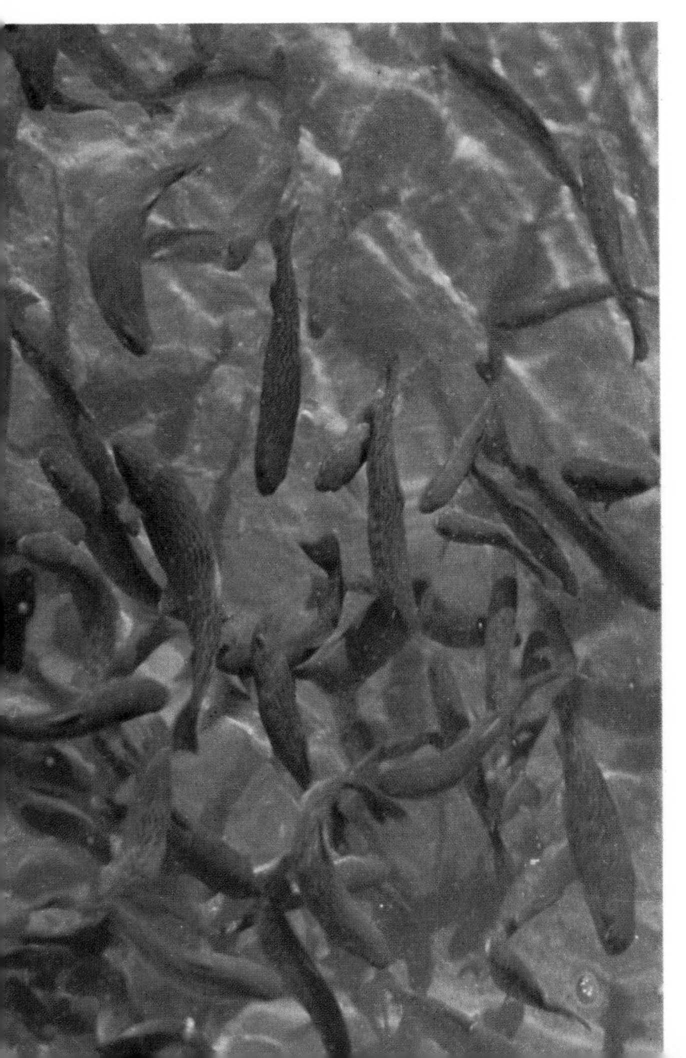

Because all living things contain water as an essential element and most of them need to drink or absorb some water fairly regularly, lakes and rivers have attracted an enormous variety of land-based life. In fact, many animals have become totally dependent on water and the life it contains. Ducks, herons, and otters are typical of this group.

And so it becomes clear that the rivers and lakes of the world constitute a major habitat supporting a complete world both above and beneath their surface. In the following pages we shall look at this amazing world from its beginnings to its possible future. And although we may often have a look at small parts of that great community, we must never forget this general perspective. For our interest in any animal or plant ultimately depends on its relationship with the total environment in which it lives. This environment is, after all, part of *our* environment, and whatever affects it is sure to affect us as well.

Life almost certainly began in the sea and it was only as a later development that animals and plants colonized the land and fresh water. However, in the hundreds of millions of years that have passed since life began, various organisms have moved from one habitat to another on numerous occasions. The freshwater fauna and flora that we see today are therefore products of several invasions from both sea and land.

The main route by which marine creatures moved into fresh water was through estuaries and thence into rivers. Some animals still make this transition within their lifetime: for instance, salmon spend years feeding in the sea, but return to breed in the headwaters of river systems, and eels breed in the sea but feed in lakes and rivers. Fish such as these illustrate some of the mechanisms that enabled animals to move into fresh water millions of years ago: their kidneys and gill mechanisms are adaptable, so that, for instance, special cells in the gills can act as either salt-secreting or salt-absorbing cells. The lampreys show even more clearly the various steps that may have occurred. Both sea and river lampreys feed on fish in the sea during their adult life, attaching themselves by suckers to the fish and feeding on their blood, but both return to fresh water to breed. During this migration, the whole excretory system changes so that they can survive the sudden decrease in salt concentration that they must face in fresh water. The brook lamprey, however, never leaves fresh water and spends its entire life in streams and rivers.

There are other ways in which the gulf between salt and fresh water has been bridged. For instance, sudden changes in land level have in the past cut off parts of the sea to create basins of salt water. This water then gradually became diluted by rainwater until the rather vague line between salt and fresh water was crossed. Any animals or plants caught in the basin either died or adapted to the changing environment and became part of the freshwater fauna or flora. It is in much this fashion that the sea lampreys that live in the Great Lakes of North America have effectively been trapped and have adapted to freshwater life.

No matter how the transition from salt to fresh water has been accomplished, changes have been gradual and relatively slight. Consequently, there is a great similarity between a freshwater fauna that has marine ancestry and related salt-water forms, not only in physical features but also in behavior such as method of feeding.

Changes necessitated by an invasion into fresh water from land, however, have been far greater. To understand why, consider the reverse process—the way in which many terrestrial plants and animals evolved from freshwater forms: early in the history of living creatures organisms entered fresh water from the sea. In their new surroundings, they would have been exposed to new hazards, such as stagnant, oxygen-deficient water or drying-up of the environment. Under these pressures some of them evolved the ability to breathe air, and some developed means of moving on land. Fins were replaced by legs, and lungs developed in the transition from fish to amphibian—to frog or newt. To return from land to water these changes had to be reversed, but animals and plants cannot exactly reproduce their ancestral forms. So, having elaborated new ways of living on land, various groups had to superimpose further changes to refit them for life in the water. Between the ancestral stock leaving water and their descendants returning, many changes occurred, and these are reflected in the many adaptations of freshwater organisms having terrestrial forefathers.

Within the different bodies of fresh water are many plants and animals that have moved in from both sea and land on many separate occasions. As their numbers have increased, so changes in the balance of the flora and fauna have occurred, and the present ecosystem has evolved. As on land and in the sea, the battle for survival has ensured that only those organisms that are best equipped to face the rigors of their environment have multiplied to form part of the existing community.

Let us look at some of the problems that an organism must face if it is to live in fresh water. Obviously those of major importance to a given plant or animal depend on its origin.

The most immediate problem is that of movement. Whether bi- or quadrupedal (two- or four-legged) it is difficult to walk in water, and it is much easier to use the limbs as paddles and to swim. However, swimming efficiently demands a number of modifications, because water must be pushed backward and down, so that the swimmer moves forward and up. Other things being equal, the faster the water is driven back, the faster the animal can swim. To move water more quickly requires either faster movement of the limbs or the development of bigger paddles. Most animals entering fresh water have adopted the latter course. As a human swimmer may wear rubber flippers, so many aquatic animals have flattened and expanded limbs that increase the surface area available to push back water. Seals, for instance, have legs modified to form large flippers, and aquatic insects have broad limbs, often fringed with stiff hairs or bristles that increase the effective area still more. Otters and many aquatic birds have skin between their toes—webbing—that greatly increases the surface area.

A second way of increasing efficiency in swimming is to reduce the drag of water flowing over the body by becoming streamlined. The perfect

Few water animals could survive a battering by Brazil's mighty Iguaçu Falls. Few sea creatures could survive even a calm river. Fresh water or certain aspects of it can be hostile to much aquatic life. Yet in almost every freshwater environment some life forms have become adapted for survival.

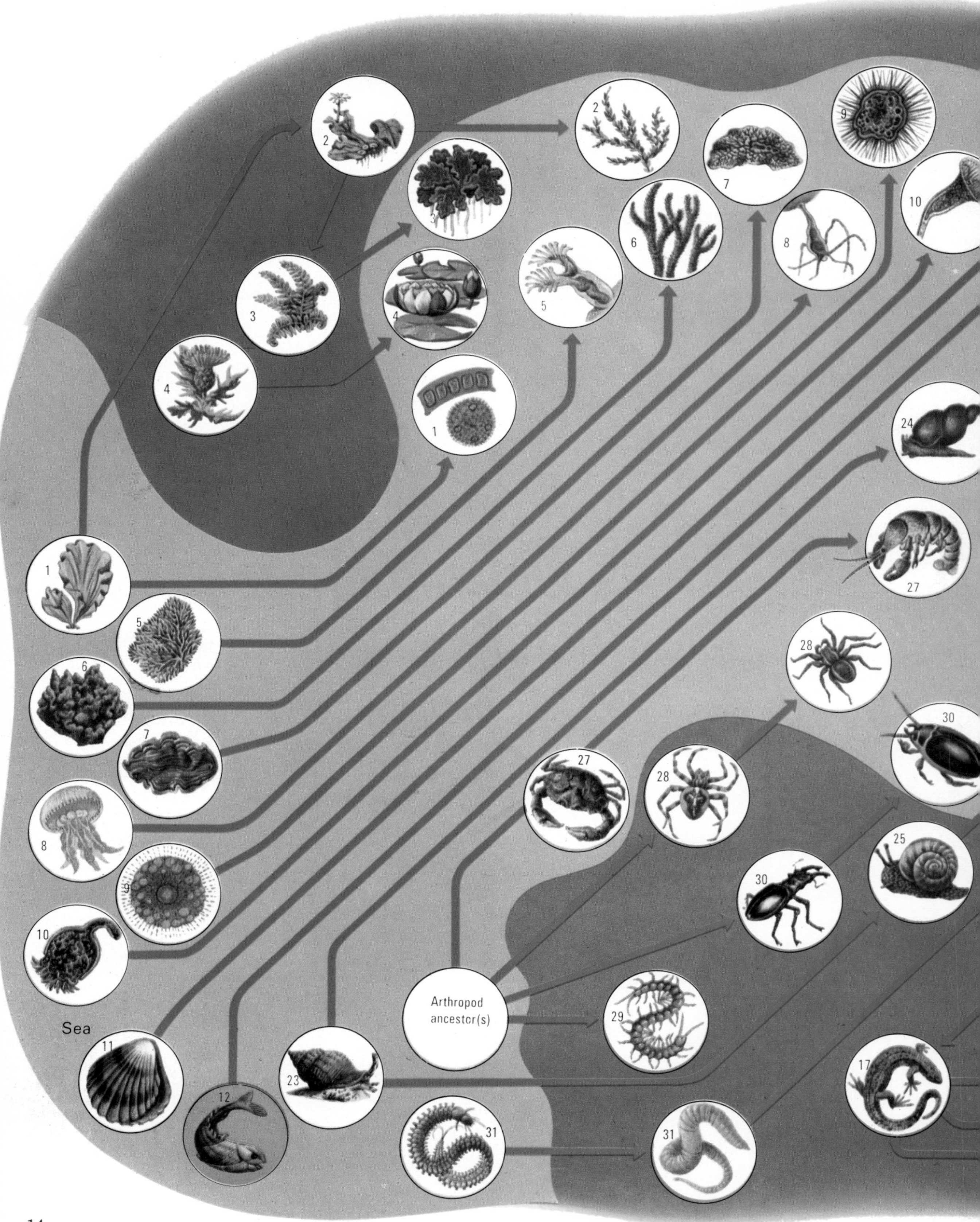

Arthropod ancestor(s)
Sea

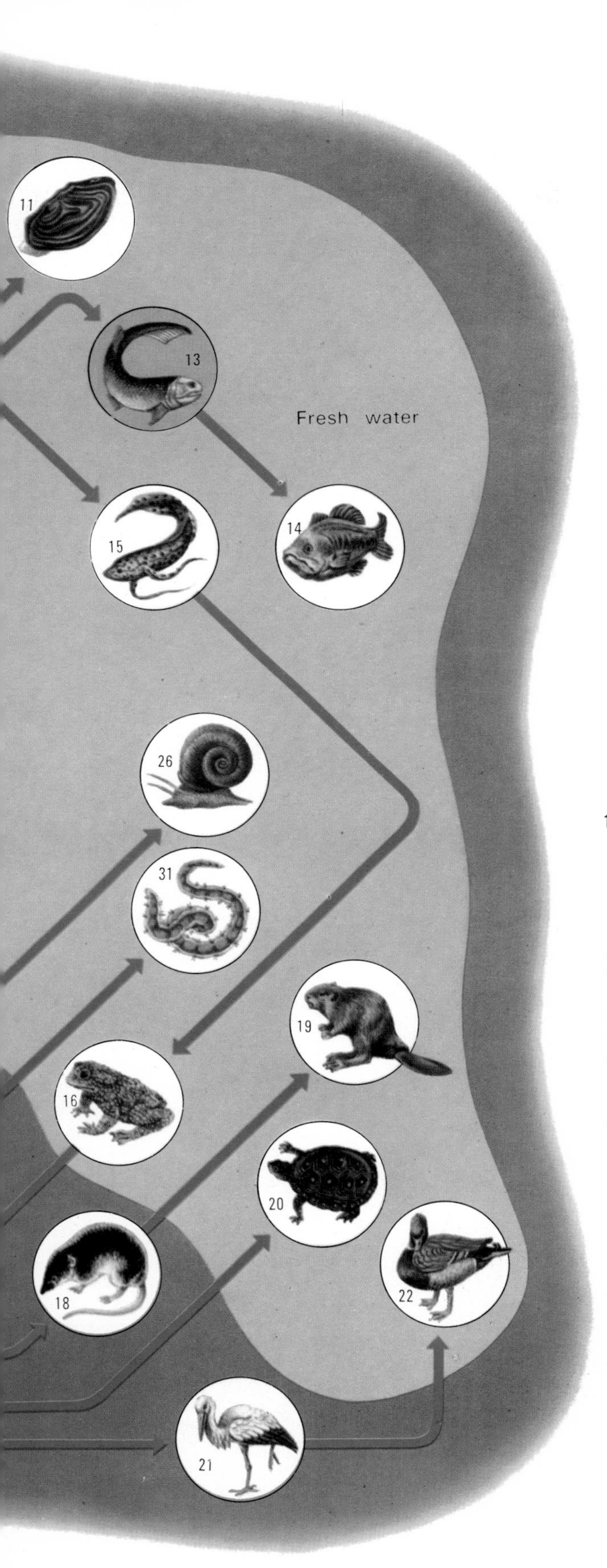

The Origin of Freshwater Organisms

1 Algae
2 Mosses and liverworts
3 Ferns
4 Flowering plants
5 Bryozoans
6 Sponges
7 Flatworms
8 Coelenterates
9 Protozoans
10 Rotifers
11 Bivalve mollusks
12 Primitive jawless fishes
13 Bony fish ancestor
14 Modern bony fish
15 Lungfishes
16 Amphibians
17 Reptiles
18 Mammals
19 Aquatic mammals
20 Aquatic reptiles
21 Birds
22 Aquatic birds
23 Snails
24 Gill-breathing snails
25 Land snails
26 Lung-breathing snails
27 Crustaceans
28 Spiders and mites
29 Myriapods
30 Insects
31 Segmented worms

This pattern of evolutionary paths shows how some kinds of plant and animal came to live in fresh water. Arrowheads show directions in which evolution occurred. We see that each freshwater organism evolved from a saltwater ancestor (or an ancestor of the saltwater organism shown). But we see, too, that different organisms reached fresh water in different ways. Many evolved from marine ancestors in a direct path from sea to estuary and river. Some freshwater organisms developed from land-dwelling descendants of marine plants and animals. But aquatic mammals, birds, and reptiles all came from land-dwelling creatures evolved from freshwater *fishes.*

streamlined shape is exemplified by some fish, which have a minimum frontal area and no large protuberances other than those used for propulsion and steering. A few of the animals that have invaded fresh water from land have been able to evolve such an efficient shape. But many animals, such as some aquatic insects, have not overcome this problem, and it has perhaps prevented more land animals from living in fresh water.

Plants do not have the problem of movement, but the associated phenomenon of buoyancy is of great significance, as leaves need to be raised off the lake bottom to reach the sunlight. Many plants living in lakes have special tissue that retains air, so their absolute density decreases and they can float. Animals, too, have adopted similar methods to make themselves more buoyant in many cases. Insects, for instance, may carry a bubble of air.

The third big problem facing an animal in water once it can move about and stay at a suitable level in the water is that of breathing. Creatures that have evolved from ancestors with gills cannot "go backward" and readopt the same breathing mechanism as their predecessors. In fact, neither mammals nor birds, which many millions of years ago evolved from fish, can breathe under the water; they must return to the surface to obtain fresh supplies of oxygen. Most insects do the same. Other groups, however, have adopted different solutions.

Water itself contains dissolved oxygen but the problem for animals is how to obtain this gas. (There is little difficulty with the breathing-out process, because carbon dioxide, the waste gas of respiration, is much more soluble than oxygen.) For a very small animal or one that requires relatively little oxygen, diffusion of the gas through a thin part of the body wall may be adequate. Many worms, for example, can obtain sufficient oxygen through their body walls. Part of a worm's body often protrudes from the mud at the bottom of the pond or stream in which it lives, and wriggles about in the water. In this manner it

The creatures that spend much time under the water have various special modifications for swimming: the cormorant (left) has a long neck, rapier bill, and webbed feet for when it plunges in search of fish near the seashore; otters (center) also have webbed feet, and

causes more water, and hence more oxygen, to flow over it. Other groups of animals have similar habits but quite often part of the body wall is especially thin and forms a "gill." In some animals, such as alder-fly larvae, the gills are conspicuous as thin sheets or filaments, but in others they are protected from damage by a hard overlying layer.

In some instances the process of diffusion may be enhanced by the presence of a respiratory pigment such as hemoglobin, the same red substance that colors human blood. Because this and similar compounds have a high affinity for oxygen, the uptake of the gas is facilitated by their presence. The pigments may also be used to transport the gas around within the body, or to store it for use in times of shortage. It is hemoglobin that gives the red color to many freshwater animals such as worms, midge larvae, which are known as bloodworms, and the ramshorn snails.

Adult insects have an outer covering that is too thick for gas to penetrate at an adequate rate. Some of them behave like mammals and get their air at the water surface. Others collect air at the surface, but instead of breathing it in immediately, take it down into the water in the form of a bubble. This bubble may act as a "physical gill," with gases being exchanged across the bubble/water interface. Eventually, however, the bubble shrinks to useless proportions and the insect must return to the water surface for a new one. A rather remarkable modification of this system is shown by some insects that retain a permanent air layer over part of their bodies. Air is trapped by numerous tiny hairs, which are hooked at the tip, so that it lies in a layer over the openings to the air-filled tubes, or tracheae, in the insect's body. So efficiently is this air trapped that it acts as a permanent "gill."

One of the most unusual means of obtaining oxygen is that exhibited by the larva of a mosquito called *Mansonia*. At the rear of its body is an extension that bears some sharp teeth and a "saw" with which the larva cuts a passage into the air-

their streamlined bodies move easily through the water; and beavers (right) have a long paddle-shaped tail, in addition to their webbed feet, to propel them strongly through the water—in this case, a sun-dappled pool in Wyoming.

Diving beetle with air bubble.

Insects and spiders living mainly under water must collect air from the surface. The diving beetle (above) traps an air bubble at its abdomen tip. The water spider (below, opposite) envelops its abdomen in a silvery air layer. Mosquito larvae (right) hang from the surface and "breathe" through special tubes. Above, opposite: unlike such creatures, fish absorb oxygen directly from the water by means of gill filaments.

Mosquito larvae below the surface.

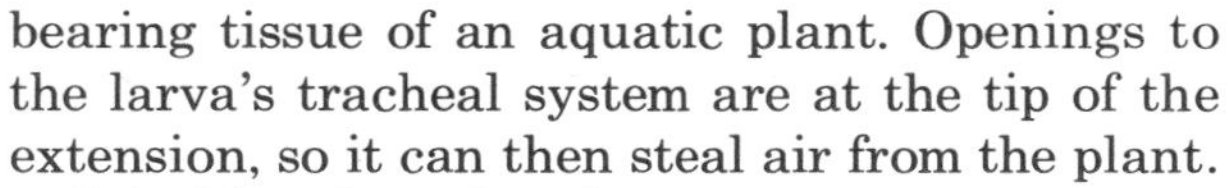

bearing tissue of an aquatic plant. Openings to the larva's tracheal system are at the tip of the extension, so it can then steal air from the plant.

Suitably adapted, using means such as those described, a terrestrial animal can move and breathe in water. But what happens if the water is covered by ice, or if it evaporates and the lake or river becomes a parched area of land?

On land, temperature changes regularly, rising and falling at dawn and dusk respectively. However, these changes are never as extreme as those that may be produced in a body of water. Water heats up and cools down much more slowly than land, so that, in the short term, the organisms that live in it are subjected to much less fluctuation than those on land, but if changes do occur they have far more wide-reaching effects than those affecting terrestrial stock. Once ice has formed, for instance, only a period of consistently warm weather can melt it. And cooler weather alone is not enough to refill the baked hollow that used to be a pond. Particularly in the latter case, many animals will die if they cannot escape to a new habitat where water remains. Some insects and birds can fly away, and some of the other creatures can crawl across damp grass or earth at night in search of an environment still containing water. If you have ever left a bucket of water outside for a few days or filled a new garden pond with water, you will be aware of the fact that aquatic organisms can move about between water bodies.

If a desiccated pond or river is refilled, however, recolonization is not totally by immigration. Some plants and animals appear to emerge as if by magic from the mud. What in fact happens is that many of the inhabitants of a dried-up area have left seeds or eggs, or have formed cocoons

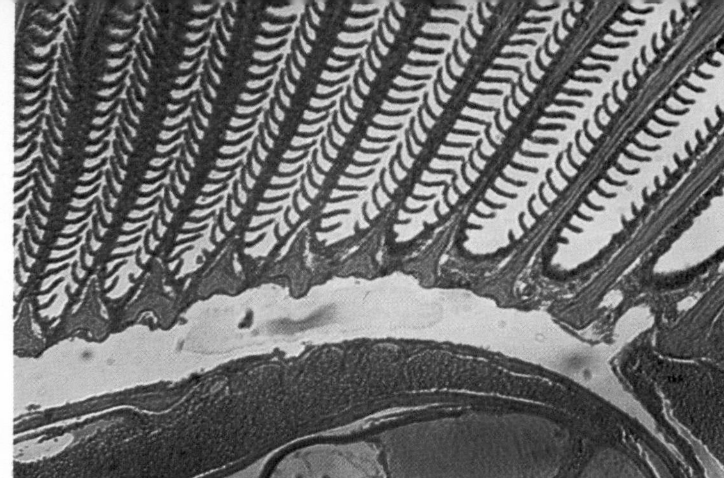

Part of a fish gill (enlarged). "Hooks" are gill filaments.

Water spider under water.

resistant to drying. Some of these, such as the eggs of the brine shrimps, are so resistant that they can be collected from the shores of certain lakes, stored in bottles for several years, and then reimmersed in water, at which time they will hatch. During a drought some lungfish build cocoons inside which they can breathe by means of their lungs. Like the brine-shrimp eggs, lungfish in this form can easily be transported and will emerge when the cocoon is rewetted.

Ice formation necessitates changes in activity that are similar in some respects to those involved in coping with desiccation. Unlike mammals and birds most animals that live in water are unable to regulate their body temperature, so their activity is limited to times when the temperature of their environment is high enough to enable them to function effectively. During periods of ice cover, they are more or less inactive. Some organisms form cocoons in which to live; others simply burrow into the mud at the lake bottom and await the coming of a thaw. Some can survive in the water beneath the ice; others overwinter as eggs—usually the most resistant stage in the life cycle. The animals that live on the water surface rather than below it may adopt the latter course, or they may leave the water altogether and move into damp areas around the edge, where they can avoid the ice.

We have said nothing so far about the difficulties faced by creatures whose ancestors came to fresh water from the sea. For them the temperature problem is the first they must face, because the sea never freezes completely, nor does it dry up. These animals may therefore adopt similar habits to those animals whose ancestors came from land. There are, however, other problems that affect only the creatures of marine ancestry.

The ability to stay alive depends on maintaining a suitable balance of chemical compounds in the body. The sea is a rich source of such substances, but fresh water is a much more dilute solution. Therefore, on entering fresh water, marine animals—whose skin is very permeable compared with the thick-skinned land animals—must find a way to get enough of the right type of compounds and to keep those that they have from seeping out. Failure to control the amounts of water and salt in the body would result in the animal swelling up as vast amounts of water were attracted into its body by the salts already there. These animals have therefore developed elaborate mechanisms to help them to obtain enough salt without being flooded. Commonly, they drink large amounts of water, extract the required salts, and then eliminate the surplus water through large excretory organs such as kidneys. Indeed, the highly efficient mammalian kidney may have evolved from that of our primitive backboned ancestors—some type of fish that lived in fresh water and had to deal with this very problem.

This question of water and "ionic balance," as it is called, is so important that it is probably the major barrier preventing more marine creatures from entering fresh water. In animals that spend only a part of their life in rivers, such as lampreys, the whole excretory system is changed during the course of their migration to equip them for the sudden decrease in salt concentration. Even animals and plants whose natural habitat is fresh water must be able to cope with different levels of salt in their environment. Some living things can cope with this variation in salt level better than others.

We often speak about hard and soft water, meaning water containing more or less of certain salts, and this alone is enough to determine whether certain animals and plants can survive. Diversity also results from minerals originating in rocks over which the water has flowed. For instance, some water accumulates so much iron-containing ore that it is colored brown (though water is often a brown color due to the presence of tannin from peat). In such waters, the inhabitants must avoid being swamped with the wrong minerals.

In summer a wealth of water lilies, irises, and other plants soften the outline of this pool in England's Epping Forest. Freshwater vegetation grows luxuriantly in favorable months in climates with marked differences between the seasons.

Two other factors are important and must be mentioned. The first is acid or alkaline content. Some water is extremely acidic, often because it has been associated with decaying organic matter, perhaps in the form of peat, whereas other water is alkaline, usually through association with chalk. Because the acidity of an organism's body must not deviate from within particular limits, the acidity of a body of water is most important in determining its suitability for a given species. Indeed, acidic and alkaline waters each have characteristic floras and faunas, which are so distinct that experts can judge the acidity of a stream fairly accurately just by examining the organisms living there.

The second, and last, important factor is light. This is of primary importance to the water plants, and it is significant that plants grow only in sunlit water. At great depths, where light does not penetrate, plants cannot survive. The actual depth of light penetration varies with the water and may be very shallow indeed if the water is murky with sediment or other suspended matter such as china clay. The plants are important to the animals as a source of food, and many animals actually live on the plants themselves, so that the distribution of animals is greatly affected by whether or not light penetration is sufficient to maintain plant growth. In other words, light may determine the spatial arrangement of entire communities of both plants and animals in fresh water.

Now that we have considered the major physical features of fresh water that affect the plants and animals, we can see why the inhabitants of lakes and rivers differ slightly from one region to another. To survive, an organism must be able to cope with every one of the problems we have discussed. If a competing organism is better equipped, its less efficient rival will die. If, even without direct competition, an organism has not become adequately adapted to the physical features of its environment, it will die anyway. Life in fresh water is a battle not only with competing forms of life but with the environment itself. Before looking at the organisms, then, think for a moment of the environment. Where does the water come from? How does it collect and flow? What does it do?

Frostbitten reeds poke up bleakly beside an icebound pool in northeast France. Harsh winters halt many activities of freshwater life, but beneath the ice life quietly persists, preparing for renewed activity with the coming spring.

Varieties of Fresh Water

Water is always present in air, usually as vapor. We are all familiar with the dewy effects of a mild morning following a cool night. Dew forms overnight when water vapor cools and condenses on exposed cold surfaces. Alternatively, the water vapor may condense in the sky as clouds. If the clouds are cooled, rain will fall. Frequently this happens when the clouds are driven up mountainsides, but there may also be rain as the result of convection—warm air rising, then cooling high in the sky. This latter situation predominates in tropical areas covered by rain forest.

Having fallen onto the ground, water flows downward under the force of gravity. In mountainous areas, rushing streams carry water on the beginning of their journey, usually to the sea. These streams, which are typically fast flowing and very cold (although some warm up quite significantly under a hot sun), emerge from marshes near a mountain top or from melting snow and ice. Or they may appear from a small hole in the ground if, beneath the surface, the water has bored a pathway through soluble rock or permeated slowly downward until, reaching an impervious layer, it has been forced to break out into the open.

These mountain streams contain few living things, and rarely any fish. The water is too cold, and it moves so fast that organisms cannot become established on the bare and rocky bottom. Once descended from the highest reaches, however, the mountain slope is slightly less steep, so the stream's flow slackens a little, and a coarse gravel or shingle forms the stream bed as it broadens out. A few small plants may cling to the banks and hang into the water. An occasional trout may be seen, and some small invertebrate creatures can be found hiding among the pebbles on the stream bed. Down and down the water flows, sometimes into a lake or pond that interrupts its course, but remorselessly down. There may be a few waterfalls or a rushing torrential stretch; the stream may disappear

The Llugwy River froths its way down a steep rocky slope at Swallows Falls in North Wales. Few water plants can gain a roothold in such swiftly flowing mountain streams, and few water animals can withstand their buffeting.

below ground or it may join other similar brooks; but ultimately the gradient lessens and the flow becomes much slower. As the force of the water slackens, its ability to carry larger stones declines, so the gravel on the stream bed becomes progressively finer. In calmer byways, a few plants grow at the stream's edge and a little fine silt accumulates around them.

The character of the stream is slowly changing. In fast-flowing water, the driving power rips away everything in its path. Any loose material that may happen to become lodged is soon pulled from the stream bed, leaving only a few bare rocks. As the flow slackens, only the smaller stones and then only silt and mud are picked up and carried along. At the same time large and then increasingly smaller particles are deposited and the rate of deposition gradually overtakes that of erosion. But as always the picture has another side: as the speed of flow decreases, the amount of oxygen in the water also decreases. In the cold headwaters the turbulence whips oxygen into the water, where it is readily dissolved owing to the low temperature. In stiller water, which is often much warmer, the amount of oxygen available for animal life is considerably less.

As the stream continues along its course it reaches the critical point where deposition starts to overtake erosion. Plants become established in increasing quantities, and the fauna is becoming more and more complex. Fish are relatively abundant. The trend toward a more complex community continues as the stream becomes a small river. The rate of flow decreases as the river widens and deepens. Silting occurs in larger amounts, and other streams enter the river. In many respects the river begins to resemble a lake, and many of the typical lake-dwelling animals do, in fact, make their homes near the edges of the main watercourse where the water is moving very slowly.

The river moves on. Ever more slowly it creeps toward the sea, depositing an increasingly large amount of silt and mud. Banks of mud accumulate and are colonized first by marsh and then by terrestrial plants. Marshy areas are frequent on the adjacent floodplain as the estuary is approached. More mud is deposited, gradually extending the mouth of the river by forming a brackish water delta. At the river mouth, except where the mud is cleared by the tide, the flow is sluggish and the water shallow. Land around the river is fertile new mud, rich in nutrients and highly prized for agriculture except where flooding is a hazard. Beyond the delta, the fresh water is caught in the prevailing tides. At first it moves as a mass, dispersing slowly. It can be detected far out in the ocean, perhaps as far as 200 miles out if the river is large—the Amazon, for example. This load of fresh water affects the marine life adjacent to its path, and so a characteristic brackish water fauna and flora prevail, derived from both the river and the sea. Eventually, however, all traces of the river are gone.

We have followed the water from the fresh mountain stream to the sluggish delta where piles of rotting vegetation, mud, and other sediments enshrine the history of its path from the hills. This is the general pattern of a river course, but the characteristic sequence may be broken in a variety of ways. Suppose, for instance, that massive earth movements gradually uplift the land over which the river has its middle course. As the land rises, the river vigorously erodes its bed, cutting its way sharply down perhaps through many hundreds of feet of rock to form the kind of narrow trench we call a gorge or canyon. The river may be wide where it enters the canyon, but the broad expanse of water must be funneled into a narrow channel, and so it churns and hurries, forced to drive even faster in the narrowest portions. The noise is deafening as millions of gallons hurtle along, tearing at the river bed, ripping out stones, crashing against the sides. At the end of the rocky walls, the river is able to spread again. It may resume a typical course, such as the one described earlier, or it may spread into an enormous thin sheet, a marsh of tremendous size, fed and drained by the river. The canyon is left behind, an impressive reminder of the power stored in the apparently tranquil waters below. The Grand Canyon in Arizona is a notable example of that power. And the enormous marsh to the north of Lake Mobuto Seso in Central Africa, which hindered the 19th-century search for the source of the Nile, is an equally remarkable example of how a river course may develop.

If we remember that fast water is powerful and erodes, and that slowly moving water deposits its

Vegetation thrives profusely on the sediments dumped by rivers along their sluggish lower reaches. Trees crowd the edge of this Amazonian river, their curved aerial roots seeking oxygen, which is largely absent from the rich but waterlogged mud.

load, rivers are to some extent predictable. This, of course, assumes that we know the type of rock over which the river flows and the volume of water that it carries. It is this last factor that can change so rapidly and with such disastrous results, in a monsoon rain or hurricane. Although we can know much about a river, we can never know *everything*, for it is never at rest. Changes can occur with such speed that controlling action will be too late. A river can be almost dry at times, a gleaming tranquil ribbon at others, and a terrifying fury at yet others.

In the course of following our stream, we noticed that it may flow into lakes or ponds. These areas of standing water may be enormous or very small, some pools being only temporary and disappearing in times of water shortage. But lakes and ponds vary not only in size, but also in shape and in the way in which they were formed.

From time to time in the history of the earth, massive changes have taken place in the superficial crust on which we live. For example, massive chunks of land have slipped relative to one another. Sometimes two parallel fractures have

Not all rivers carve steadily broadening valleys: the Colorado River flows for 280 miles through a narrow gorge a mile deep, creating Arizona's famous Grand Canyon.

occurred, and the ground in between the fractures has sunk down, thus forming a rift valley. These valleys often contain lakes whose borders are bounded by the two fracture faces. Examples of such lakes, which are long, narrow, and deep, are Lakes Tanganyika and Malawi, which form part of the chain of lakes occupying the famous rift valley of eastern Africa. The lakes are fed by streams off the adjacent land blocks, and their outfall is along the valley floor. The combination of deep lake and high neighboring land masses gives rise to the spectacular scenery that is so characteristic of this part of eastern Africa.

Volcanic activity is another cause of changes in the surface of the earth's crust that may result in the formation of lakes. Volcanoes are formed by the expulsion of *magma*—molten rock material—from a chamber in the earth, and whenever the walls and floor of the central vent, or crater, of the volcano lack drainage channels, it may be filled by a lake. Volcanic crater lakes often have an almost circular outline and they may be very deep. The existence of volcanic lakes has been realized for many years—for instance, Lake Viti

Rivers that reach obstacles or hollows may form lakes. This cormorant-haunted lake fills one of the depressions in the East African section of the great rift valley.

in Iceland (one of the first actually to be recorded) resulted from an eruption on May 17, 1724.

Less spectacular in their formation but more numerous are the lakes that resulted from glacial activity far outside the present polar regions during the ice ages. The most famous glacial lakes are those that formed when glaciers slid down valleys in mountainous regions, deepening and widening the valleys as they went. During its journey such a glacier collected rock debris, or *moraine*, which was dumped as it reached the melting front of the glacier at the mouth of the valley. The moraine acted as a natural dam, and the basin behind it filled with water, thus forming a deep lake. Such lakes are usually long and narrow because they largely lie in the valleys that were once occupied by the glaciers. Scandinavia, northwest England, Scotland, and the Alps abound with these lakes; and the famous Finger Lakes of New York State were somewhat similarly formed by tongues of ice.

Glacial action has left numerous other types of lake. Some, like those in Wisconsin, resulted from the melting of stationary glaciers on flat land.

These are much shallower than the lakes resulting from moving glaciers, and they have an irregular outline circumscribed by debris deposited by the melting ice. Much smaller lakes in hilly country may be due to the action of water that, in successively freezing and thawing, fractured and crumbled the surrounding rocks. Cirques or corries, which are formed when glaciers enlarge and deepen hollows at the heads of valleys high up on mountain slopes, also often contain lakes.

Even in the period since the last ice age, the power of running water has reshaped many landscapes. And in certain circumstances lakes may be created in the course of a river. For instance, the spectacular Grand Coulee area in Washington State was once in the course of a river. Impressive waterfalls then cascaded into the valley, cutting deeply into the rock at the base of the falls. This carved out the pit, now known as Falls Lake despite the absence of the cascades.

Other well-known lakes derived from the action of running water are the *oxbow* or *horseshoe* lakes, such as those associated with the Mississippi, or with the Darling and Murray rivers in

Wastwater in England's Lake District lies in a valley scoured to below sea level by an ancient glacier, and dammed by glacier-borne rock debris dropped where the glacier melted. Many glacial lakes originated in this twofold manner.

Australia. These lakes, so named because of their shape, are due to the processes of erosion and deposition that were mentioned above in our discussion of rivers. On a flat plain composed of soft ground, if a river is even slightly deflected from a straight course, the bank on the outside of the bend will be eroded, while that on the inner side is subject to deposition. In the long term, this results in a series of bends, or *meanders*, some of which may subsequently become isolated from the river and form oxbow lakes. The primary reason for this isolation is the powerful horizontal cutting action of rivers, intensified in times of flooding, so that a river sometimes cuts through the neck of a bend.

Another major group of lakes are those that have occurred as a result of the solution of rock. Limestone and chalk rocks are composed of calcium carbonate, which reacts with the carbon dioxide dissolved in rainwater to produce soluble salts. This soluble matter is washed away, and as this process continues, a depression of gradually increasing size is produced. If the underlying rock is impervious this hole may fill with water, so becoming a lake. Deep Lake in Florida is an almost perfectly round *solution lake*, as such lakes are called, but more irregular shapes occur in the Alps where glacial action has also been involved. Lakes may even be found underground in caves in limestone country. The caves are formed by solution by underground streams in a similar manner to that in which rainwater results in solution lakes above ground.

Although we have noted many different types of lake, not every body of water falls into one of these categories. Lakes are also found in old meteoritic craters, behind dunes of debris built up by the wind, behind masses of vegetation, enclosed by sand dunes near the coast, and even behind beaver dams. Often two or more factors have been involved in producing a single lake.

These, then, are some of the natural types of lake found on earth. Man himself has added enormously to the list by his activities. Excavation of minerals and gravel has left holes that have then filled with water. The value of such man-made lakes has increased markedly in areas of high population density where the amenity

Aerial photography reveals an oxbow lake beginning to form where a meandering stream joins Utah's Bear River. Oxbows form on flood plains, where rivers frequently change their course.

aspects of water are prized. Damming of rivers, whether for the irrigation of adjoining land, for power from hydroelectric generators in the dam wall, or for a supplementary water supply, has created many new lakes or reservoirs in the last 100 years or so. The Aswan Dam in Egypt, the Tennessee Valley Project, and Rhodesia's Kariba Dam are among the most widely publicized of such undertakings in recent times. Many smaller ponds have been man-made, too, and man has created them for a multiplicity of purposes, especially back in the days when life was more widely agricultural. Watering ponds for livestock are an obvious example.

One other familiar type of standing water requires a word or two: the marsh. Marshy areas are usually due to flooding by a river or by the sea. In the latter case, silt may build up and trap the water over an impenetrable layer, such as clay. This may result in the growth of an area of lush vegetation, which utilizes the massive influx of salts and minerals. Dead material may accumulate as peat and lead to the formation of low-lying land, such as that in East Anglia in England. Marshes that are not caused by flooding by

Above: Lake Powell in Utah and Arizona is an entirely man-made lake more than 100 miles long. It consists of the waters of the Colorado and tributary rivers penned back by the Glen Canyon Dam in Arizona.

Left: surface water appears as bogs and marshes as well as lakes and rivers. This Welsh bog's standing pools flanked by cotton grass and pale green bog moss typify the poor natural drainage to be found on large tracts of rain-soaked upland.

surface waters may be the result of water being permanently present just below ground level, often because of the proximity of a lake or river.

We have taken a rather long look at different types of lake, river, pond, and marsh. Why? Because the physical characteristics of any body of fresh water, whether standing or running, depend on where it is, on the rock around and beneath it, and on the land adjacent to it; and because the animals and plants living in the water depend for their very existence upon just these same physical characteristics.

Bodies of fresh water can be grouped in many ways on a biological basis. The most common methods are based on an assessment of productivity: how many and how large are the organisms, particularly algae, in the water? From time to time, fairly strict categories have been set up; but more recently we have begun to avoid rigid forms of classification because of our increased realization that lakes and ponds change biologically, and can move from one category to the next. But though we now recognize a series of very fine gradations, some former categories still have a useful applicability. Take, for instance, the terms *oligotrophic* and *eutrophic*.

Water draining into lakes off hard rocks that bear little vegetation contains rather few minerals. As life in the water depends on this input, any lake fed by such water is relatively unproductive; in other words, there are few animals to use the oxygen that is therefore always available at the bottom, and the lake is said to be oligotrophic (from the Greek *oligos* meaning "small" and *trophe* meaning "nutrition"). In contrast, eutrophic lakes are productive: they have a rich flora and fauna, which may deplete the oxygen supply in the lake bottom during the summer. Eutrophic lakes are those situated near farmlands or areas of lush vegetation, where the drainage into the lake is consequently rich in minerals and organic matter. They are found typically in lowland areas with soft, sloping rocks from which minerals are dissolved by the slow passage of water.

As has been said, however, lakes can change, and cases are known of productive lakes becom-

ing unproductive because of the slow loss of nutrients from the water. The converse is also possible, and oligotrophic lakes may become eutrophic, frequently as a result of the actions of men. For instance, the use of fertilizers for the artificial stimulation of plant growth may result in the enrichment of water draining into a nearby lake, and therefore the nutrients available to its plant and animal inhabitants are increased.

So we have seen that rivers and lakes show immense variety. From the mountain stream to the river delta to the eutrophic lake with no oxygen in the deeper parts, there is a whole succession of different environments, most of which provide a home for at least a few animals and plants. We have looked at fresh water and some of its properties, noting the problems of living in it, and its power to change the landscape. Now we shall turn our full attention to the animals and plants that have won the battle for existence in and around the water. Let us begin with the success story of a lake. We shall consider the richest freshwater environment and see how the animals and plants interrelate in the world under the water surface, how they live, and what effect they have on each other.

Two freshwater localities with very different potentials for supporting life. Left: a farm pond near Napier, New Zealand. Organic nutrients washed from surrounding pastures support a dense growth of duckweed. Above: a stream near Kinlochewe in northwest Scotland. Rocks flanking the stream yield few nutrients, and these waters will be largely sterile.

The Features of a River System

This imaginary landscape tells the story of a river from source to mouth. The river rises as a swift mountain stream (top right). The stream plunges down into a glacier-scoured hanging valley to form a small mountain lake. Water from this lake and other sources fills a much larger lake with an underground outlet through a cliff face. This underground river flows through a cavern (shown inset) worn by water through a mass of limestone. The river emerges as a waterfall cascading into a canyon to join another river. The combined waterway broadens as tributaries join it, and the steep-sided spurs that hemmed it in give way to the low banks of a flat floodplain where an oxbow lake marks an old abandoned stretch of watercourse. Finally, the river forms a tidal estuary where mud islands have begun to build a delta out into the sea.

The Lake: Cycles of Life

The sun is slowly falling. It has been a beautiful day, and the water lies calm. A few birds still sing in the wood opposite the place where we stand. Cattle from the adjoining field cool themselves, standing at the water's edge. For some time the days have been shortening; the trees are golden brown, and leaves carpet the ground. The autumn breeze has lazily blown some leaves onto the water surface, where they drift lifelessly. A fish rises to take a fly, and the "plop" breaks the silence over the lake.

This is the magic of nature. It is the uniform peace, along with the strange feeling that all the things we see "fit together," that has created our illusion of a calmly quiescent scene. It really does not matter where we are. This temperate scene could be in North America or England or Australia; it is a picture that we can all share.

But let us break the spell and look in more detail at what is happening. In reality, this lovely autumn evening represents the climax of many events that we have already considered. The lake is eutrophic, fed by the water running off the adjacent land. This water flow carries products of the wood, many fertilizers from the farmland, and, of course, animal wastes. The continual replenishment of nutrients in the water has encouraged the lush growth of plants, and this in turn has enabled a wide variety of animal life to become established.

The lake is very complex. It provides many homes of different sizes. Some fish are moving around almost the entire body of water. To other creatures, a single reed or even a decaying leaf is the whole world: they live, breed, and die in this small space. What happens elsewhere in the lake is unimportant to them except as it affects their minute dwelling place. Such tiny situations are called *microhabitats* and hundreds of them are to be found in any lake. They merge indefinably into more extensive environments at the upper end of the scale and even tinier habitats at the other.

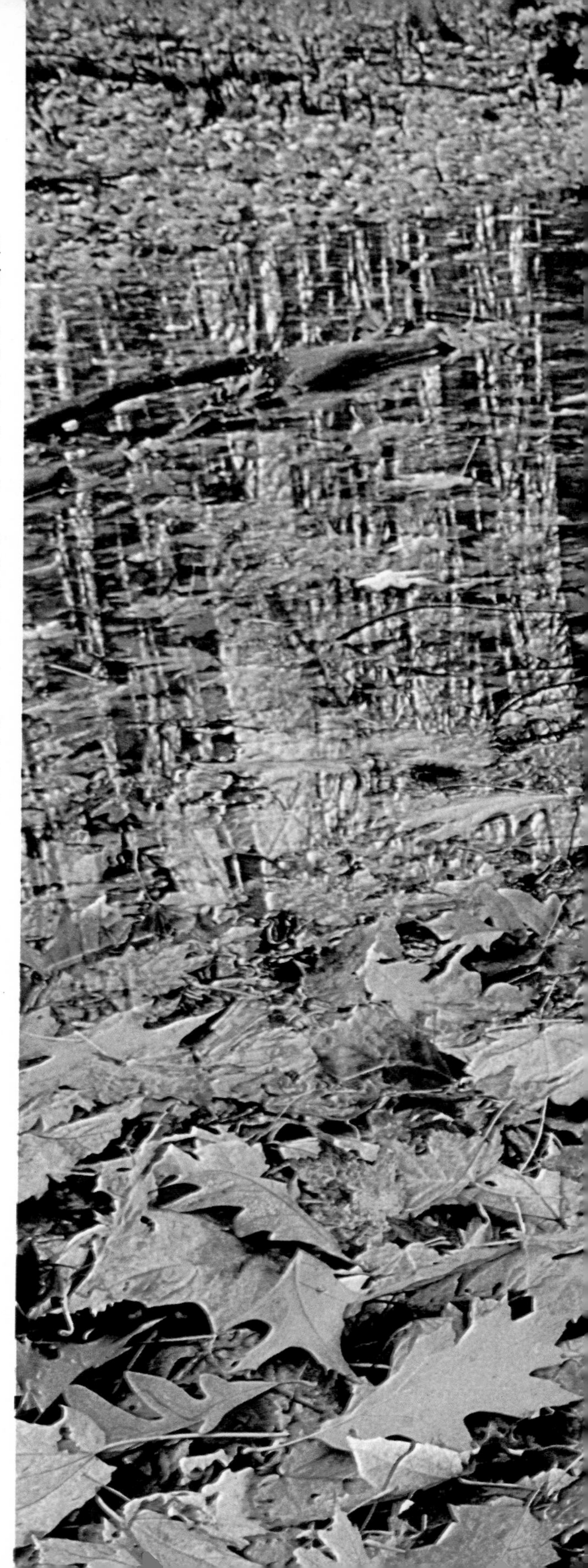

Fallen leaves make pleasing autumn patterns in a woodland pool. The rotted leaves and plant and animal wastes washed into the water will nourish tiny plants and animals—lowly members of the pond's food pyramid. Thus the nutrients upon which freshwater organisms depend are constantly recycled.

By way of introduction, let us return to the world with which we are more familiar—the land. All the elements necessary to maintain life move in endless cycles. Carbon, sulfur, phosphorus, and many others circulate through different living organisms. If circulation stopped, each element would accumulate in some stage and be lost to the rest of the system, resulting ultimately in a desperate shortage and consequent death for other organisms. Each element has a rather different cycle, although they may overlap in some stages. Let us take a very simplified example of what happens. Grass contains all three of the elements mentioned, combined in substances called *proteins*. Grass is eaten by rabbits, which utilize its elements, regrouping them to produce new kinds of proteins. In the rabbits' feces and urine the excess of the elements is voided to the soil. If a rabbit is eaten by a predator, the elements move on again; some are lost as before, and others are regrouped again as proteins. Eventually the predator dies, its corpse rots away, and the elements—now in yet other altered forms—join those in the soil derived from animal waste and dead plants. After various stages in the soil, the elements are reabsorbed by the roots of the grass, and the cycle starts again.

Additional factors may complicate the cycle but do not disrupt it. The grass photosynthesizes—in simple terms, makes sugar from water and carbon dioxide in the air—when it is exposed to light. Thus, some carbon joins the cycle at this

Above: the joining of the muddy Kawarau River and the clear Clutha River in southern New Zealand hints at differences in the food cycles of the lands they flow through. Overgrazing has stopped grass regrowing on the Kawarau's valley slopes, baring the valuable topsoil so that rain washes it down into the water. Not only does this disrupt a land food cycle, it may also damage life in the affected river.

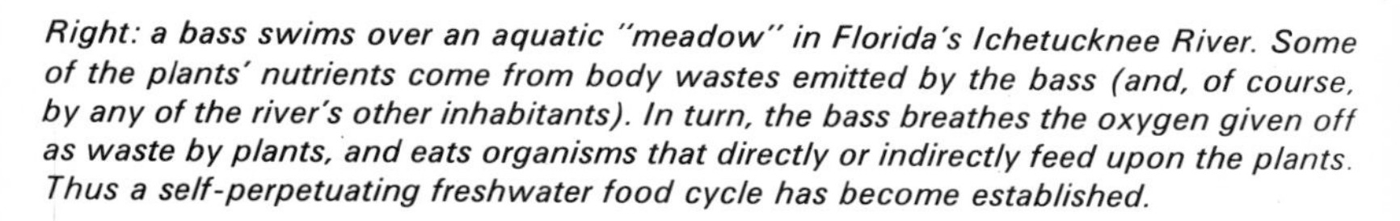

Right: a bass swims over an aquatic "meadow" in Florida's Ichetucknee River. Some of the plants' nutrients come from body wastes emitted by the bass (and, of course, by any of the river's other inhabitants). In turn, the bass breathes the oxygen given off as waste by plants, and eats organisms that directly or indirectly feed upon the plants. Thus a self-perpetuating freshwater food cycle has become established.

point, and may return to the air from animals, which breathe in oxygen and emit carbon dioxide. Cycles are thus superimposed until remarkably diverse patterns, involving hundreds of steps, are created. What is more, the cycles are naturally self-regulating. Changes in the environment result in modification and emphasis of one pathway in the cycle over another. But the system is not upset as a new equilibrium is soon established. The natural balance is preserved and disasters can be averted—except when, as we shall see later, man thoughtlessly intervenes.

The steps in the terrestrial cycle are generally familiar. In lakes, although the animals and plants are not as well-known, the pattern is nevertheless the same. To see how a lake works and how its living organisms interrelate, let us follow a similar cycle to the one we have just sketched.

Autumn leaves are drifting onto the surface of our lake. Night falls and a breeze stirs the water into gentle motion. Ripples chase each other across the water, the air is damp, and drizzle pits the water's surface. The water swirls below, churning its contents into slow motion. The dead leaves, driven along, begin to sink slowly. Distributed by currents, they come to rest in heaps on the muddy bottom, where they are bedded and compacted by drifting silt whipped up by the moving water. The stratification of water that persisted during the summer is now broken. For some weeks the bottom layer was devoid of oxygen, unstirred by wind, rain, or

current. Now life returns in quantity. Activity recommences with the return of oxygen and leaves. The temperature, which had fallen below that of the surface waters, begins to rise as the water is mixed. The *thermocline* (the dividing zone between the oxygen-rich upper layer of warm water and the cold, oxygen-depleted lower regions) is disrupted. The two zones—the upper *epilimnion* and the lower *hypolimnion*—were separated during the summer months by the marked differences in density, the warmer water floating on the cold below. Little mixing occurred between them, and the differences were magnified as the water in the upper zone gradually got warmer in the summer sun.

Although more and more leaves are entering the water, they do not accumulate long enough to fill the lake. The lower layers are continually decomposing, that is, being broken down into their component elements by the activity of microorganisms, including bacteria and fungi. As in the soil on the land, the mud at the lake bottom is rich in such microorganisms—living things too small to be seen with the naked eye. This invisible world is of prime importance; other life would be impossible without it.

Bacteria usually occur singly, though some form groups or threadlike filaments. Most creatures are composed of many subunits (cells), but the bacterium is merely a bag of material, less complex than even a simple cell. Many bacteria depend on water currents to move them about but a few have a thread or *flagellum* that they can beat to propel themselves. All bacteria lack the green pigments of plants, and most feed on dead organisms, whether plant or animal, although a few are parasites, and feed on living tissue. They all eat by releasing chemicals called *enzymes* that digest their food outside their bodies, and the products are then absorbed. Some bacteria require oxygen like all other creatures, but some can live without it. Thus bacteria of one type or another are able to live in many situations and certainly in all natural waters. In addition, in adverse conditions they may produce hard, resistant spores that enable them to survive until suitable conditions return and new bacteria can develop with some chance of survival.

Bacteria in fresh water are of many kinds. In the process of obtaining their own energy supplies they play various roles in decomposition. Some are involved in the initial digestion of organic remains and yield ammonia as waste. Others use ammonia to build new compounds such as nitrates. Others produce the sulfureted hydrogen, smelling like rotten eggs, that sometimes emanates from stirred mud. And still others are responsible for making marsh gas, an explosive mixture that sometimes catches fire.

Some bacteria found in water are normally terrestrial but get washed into the lake by streams and other forms of "runoff." Others are typical inhabitants of the guts of animals, including man, and their abundance in lake water serves as an indication of the amount of sewage present in the lake.

Besides the numerous bacteria that are involved in the initial decomposition of leaves, fungi also play an important role. Unlike other plants, fungi do not contain the green pigment chlorophyll, so they cannot manufacture their own food by photosynthesis, and mostly feed by breaking down dead matter. Most people are familiar with mushrooms, yeast, and molds (such as *Penicillium*) that grow on old bread and cheese. These are but a few of hundreds of different kinds of fungus. Those in fresh water are known collectively as water molds, and play a most significant role in decomposing dead matter, both by mechanically breaking it up into small pieces for consumption by other creatures, and by chemically turning some into its constituent parts. They specialize in hard objects, such as thick pieces of wood and the shells or skeletons of insects and crustaceans; and in large numbers they may look like a furry coating on an affected object. Like bacteria, a few fungi feed on living tissue: for example, a furry coating can often be found on wounds on a fish's body.

Under a microscope the fungi appear as long threads, or *hyphae*, with occasional darker lumps on the ends of some filaments. Such lumps release thin-walled, free-swimming spores into the water or they may produce thick-walled cysts that can resist extremes of drought and heat. All fungi require oxygen.

These two groups, the bacteria and the fungi, are the most important factors in the initial decomposition of leaves and other debris on the lake floor. Their activity breaks down the material to a state in which other creatures can take over the process of decomposition. Among the most important of these are some of the single-cell organisms known as *protozoans,* which resemble certain kinds of algae and may well have a common ancestry with them. Most

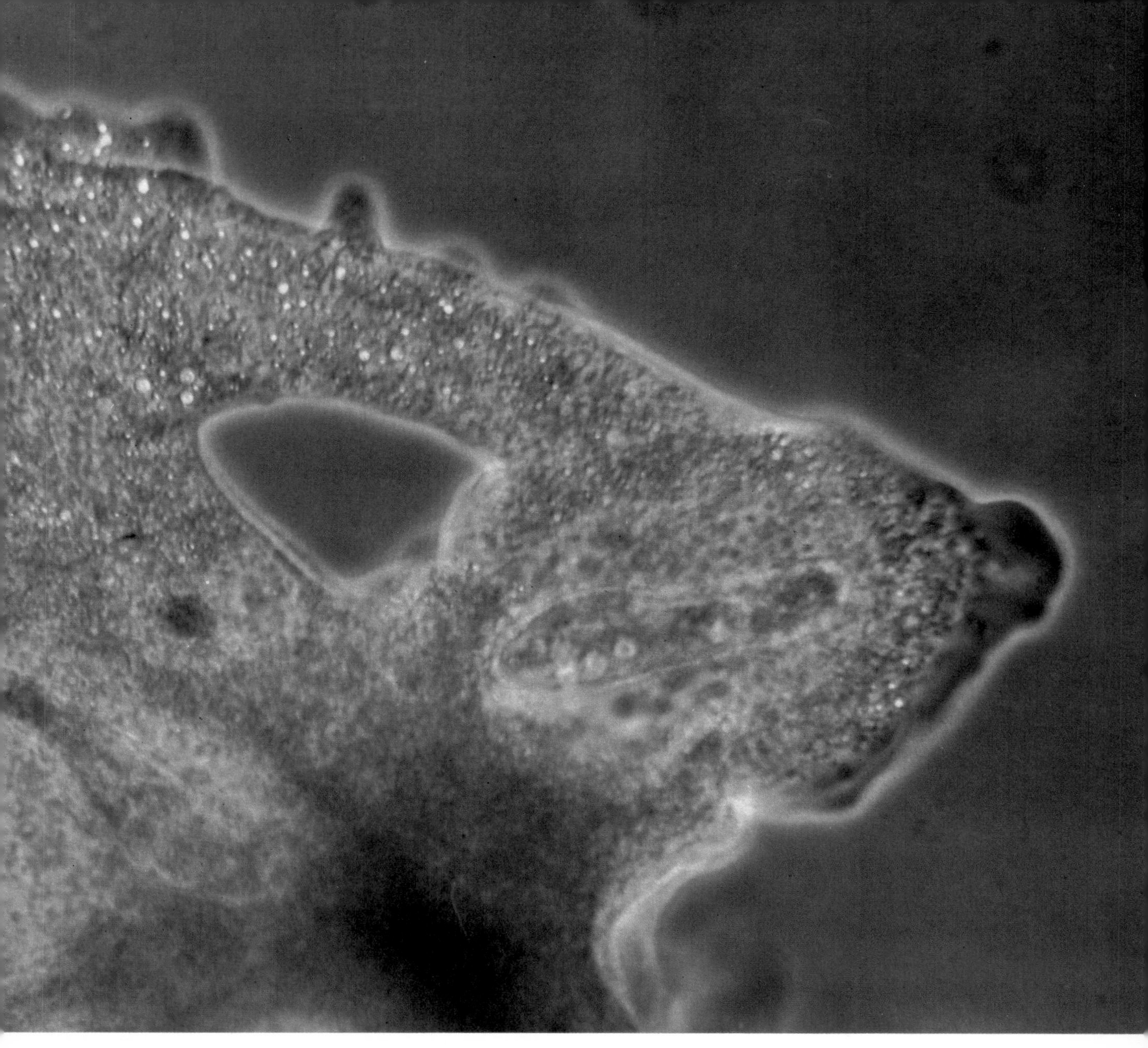

A highly magnified view of part of an amoeba absorbing a slipper-shaped paramecium (right). After bacteria and fungi, amoebas rank high among freshwater scavengers.

protozoans lack chlorophyll, although some members of one group, the flagellates, do contain the green pigment and may therefore be considered as either animals or plants. The most important protozoans concerned in decomposition are a group called the rhizopods. In this assemblage is the well-known amoeba, which looks like a tiny mass of gray jelly. Like its relatives, it moves by pushing out "limbs" of jelly, called *pseudopodia*, and then flowing along until the rest of the body catches up.

Amoebas can be found gliding over decaying leaves or mud. They engulf their food by means of a pseudopodium, which surrounds the item of food and then fuses beyond it. Imagine, for instance, pushing a marble into a lightly inflated balloon and the balloon then enclosing the marble and separating it from your fingers. The marble would then be inside the balloon. This is how an amoeba takes in its food. When the food has been digested and absorbed, the amoeba leaves its waste behind by simply reversing the

Scooped from their native mud, Tubifex *worms are sold for petfish food in writhing balled masses like this one (much enlarged). These worms can feed on urban river sewage in conditions that would poison or suffocate most kinds of fish.*

procedure. Anything of suitable size may be engulfed: for example, bacteria, tiny algae, small pieces of organic debris, or a piece of dead leaf.

Although amoebas are masses of simple living protoplasm, related rhizopod protozoans usually have shells. Some secrete the entire shell themselves and look as if they are living in miniature inverted bowls, but others secrete a very thin shell that forms the basis for a thicker layer of material, often sand grains or pieces of leaf, collected by the animal. These shelled forms protrude the pseudopodia from a single aperture. Other rhizopods include the attractively shaped sun animalcules, which have a great number of fine pseudopodia radiating from a central mass of protoplasm. These sun animalcules never retract the pseudopodia, as amoebas do in order to move, but roll along with just the tips of the pseudopodia touching the substratum. Like other rhizopods, however, the sun animalcules do use their pseudopodia to engulf their food. Although rhizopods are typically found crawling on solid objects, some are found free-living in plankton.

We discover a rather different life style among the ciliated protozoans. The body of most of these is covered by tiny flickering hairlike projections, or *cilia*, which move in waves to propel the animal along in open water. Food is ingested through a definite mouth and passes along a predetermined pathway before the waste is voided. Many ciliates feed by using their cilia to create a current that drives food particles into the mouth—a technique called *filter feeding*. They consume a wide range of small creatures as well as vegetable remains. Some are on stalks, and they may form balls, or colonies, that can be easily seen with the naked eye. For instance, the bell animalcules are typical stalked forms, with a conical or bell-shaped body supported at the pointed end by the stalk. The stalk can be contracted, but when fully extended it may be as much as three times the length of the body.

There are other types of protozoans in our lake, but those mentioned so far play the most important part, along with the bacteria and fungi, in turning dead leaves and other organic matter back into living tissue. Microorganisms are not, however, the only inhabitants of the mud on the lake floor. Other, larger animals also feed on the vegetable debris. To take a look at them we must move a few steps up the size scale.

Among the most important animals involved in the decomposition of vegetable remains are worms. As in the soil of gardens or fields, worms in the water are far more numerous than many people realize. There are many different kinds of worms in the mud of the lake bottom. They all

have a tough but supple outer skin, or *cuticle*, and a body composed of many clearly delimited segments. The skin is provided with bristles, or *chaetae*; but unlike their marine relatives, fresh-water worms have relatively few of these. The most common worms are not more than an inch long, and possibly the best-known of all are those in the family Tubificidae. *Tubifex* worms, frequently fed to fish in aquariums, belong to this family. Normally, they live in the muddy bottom of lakes or rivers, with their heads in the mud and their tails waving about in the water. Their bodies contain the red pigment hemoglobin, and where there are large numbers of them, they may give the mud surface a reddish hue. With their hemoglobin, and their swinging tails, which circulate the water around their bodies, they can withstand conditions in which oxygen is available only in small amounts.

Other worms found in the mud include the long-

Left: a female starhead minnow inspects the unusual freshwater shrimp, Mysis relicta, *in a Georgia stream. In the past this shrimp's saltwater ancestors were cut off from the sea in land-locked water gradually made fresh by rain and inflowing rivers. Unlike most sea creatures, this species of saltwater shrimp slowly adapted to the gradual change to fresh water.*

Below: these scavenging shrimps, called gammarids, thrive in streams and ponds over much of the world; they swim, leap, and breathe with the help of three kinds of limbs that project from their bowed, and rather flattened bodies.

bodied species in another family, the Lumbriculidae. Some have red pigment like *Tubifex*, which may be obscured in parts by a green pigment in the body wall. Others are translucent and have no pigment. Many of these worms reproduce by simply breaking up into pieces, each of which grows a new head and tail.

Some other families of worms show equally peculiar means of reproduction. The family Naididae, which is probably the largest freshwater group, is one example. Individuals occur in chains and new members are budded off at the back. Very occasionally, though, they may lay eggs. Worms of this family show great variety in both body form and habitat. Some live in shells or tubes vacated by their previous owners, some construct a rough and flimsy tube of their own, and some live free in the mud, without a shelter. One type of Naididae has even reversed the *Tubifex* habit of head in the mud, tail in the water: a special gill area permits it to keep its head out of the mud, leaving its tail anchored.

And there are still other families enormously different in form from those we have looked at. Some are large and glossy blue, others tiny and translucent, and others very long and thin. Species in the same family as the common earthworm (the Lumbricidae) are rare in fresh water, but those that do occur bear a close resemblance to their land-dwelling relatives. The pot-worms (Enchytraeidae) are very common, particularly around the lake margin, where they live among the roots of plants. They are usually small and white and, like *Tubifex*, are familiar to aquarists as fish food.

Although there are many different kinds of worm living in and around the mud and decaying leaves of the lake bottom, they do not vary much according to geographical distribution. True, there are slight differences between the worm populations of certain lakes in the different continents, yet in all but the most extreme cases, similarities are greater than differences. The same may be said of many other groups of freshwater animals and plants, including freshwater shrimps, which also play a major role in the decomposition of vegetable debris.

Most freshwater shrimps are found in oxygen-rich rivers and streams, but some also occur in lakes, particularly those that are not very deep. Though they feed mostly on vegetable matter, they are in fact omnivorous and may eat small animals, including their own young. Closely related to the beach fleas or sandhoppers often seen near rotting seaweed on the beach, they have flattened, narrow bodies, with a reflexed abdomen that curls under the head and thorax (the middle part of the body). In color they vary from gray through red to almost yellow, and they usually measure no more than about an inch in length. Many of the species look very similar, and the similarity persists for the most part no matter where in the world you find them. One group of these shrimps, the gammarids, have front limbs that are flattened and are used as paddles for swimming; the rear limbs are used for jumping, and those in between produce a water current over the gills, which are situated at the bases of the thoracic limbs.

Related to the freshwater shrimps are the water lice, also known as sow bugs or hog slaters. Unlike the shrimps, they are flattened from top to bottom and are crawlers rather than swimmers. In fact, they are closely related to wood lice, and certainly they look very similar. As with wood lice, their staple diet consists of rotting vegetation. The most common slaters are those belonging to the family Asellidae. They are usually brown or gray in color and measure up to an inch in length.

Water lice can easily be kept in a jar or dish of water with a few dead leaves. If you collect some females in late spring, you can see the eggs or young water lice and it is fascinating to watch the tiny transparent youngsters growing up and developing adult coloration after they have left the brood pouch.

There is one other group that constitutes a large part of the community living in the mud at the lake bottom. This group is, of course, the insects—primarily, in this habitat, insect larvae. Because of the immense variety of such creatures, we cannot generalize meaningfully about them. So we will examine just a very few examples of saprophytic freshwater insects—those that feed on dead and decaying organic matter.

The most important adult saprophytic insects in fresh water are the water boatmen of the family Corixidae. These are water bugs, in the true sense of the word "bug," although they have a number of aberrant features. They are unlike other bugs in that they do not get their food by means of piercing mouthparts that penetrate the tissues of plants or animals. Instead, they suck up all sorts of small particles from the lake floor, no doubt including not only decayed matter but also a selection of living algae, protozoans, and diatoms. In appearance these small bugs rarely exceed half an inch in length, and they generally look alike even though there are actually many kinds living in different types of fresh water. They are usually black or brown, with hardened forewings folded over the abdomen, which may bear transverse yellow stripes. The visible upper part of the thorax may bear similar markings, and the head is triangular when viewed from the front. As in all typical insects, there are three pairs of legs. The hind pair, the only one used for locomotion, serves as oars in swimming, and the surface area is broadened by means of long fringed hairs. The middle pair of legs is used to hold on to plants at the bottom of the water, and the front pair, which is much smaller than the other two pairs, is used by the male for *stridulating*—making shrill creaking noises to the female during courtship. Some adult water bugs can fly, using the delicate hindwings that are normally hidden beneath the toughened protective upper pair of wings. The young closely resemble their parents but, like all immature insects, they lack wings.

Among larval insects involved in the decomposition of plant debris, the most important are those of the nonbiting midges. Although there are many kinds of nonbiting midges living in a wide variety of habitats, they all have a similar life history. As an example, let us take the most familiar kind, whose bright red larva is sometimes called a "bloodworm." The adult female midge lays her eggs in a gelatinous mass at the water surface, sometimes anchoring it to a suitable object—a leaf or twig. Very small transparent larvae hatch from these eggs and swim down to the mud on the lake bottom. The larvae, which cast their skin three times, develop the characteristic red color after the second molt. Then, after the third molt, each individual metamorphoses into a pupa, which has a crown of fine white filaments and is a common model for fishermen's lures. Eventually the pupa leaves its tube, which was originally built by the larva in the mud, and rises to the water surface, where the adult midge emerges. Clouds or swarms of male midges can often be seen flying about on fine evenings. These swarms have occasionally given rise to false fire alarms because they keep changing shape in the prevailing wind, just as smoke does. Although the females enter the swarms to mate, they generally do not remain in the cloud of males for very long.

But it is their larvae, not the midges themselves, that play an important role in the decay of organic waste. The larvae use their saliva, or silk, to build tubes in the mud to which small particles may adhere. They live in these tubes, and feed either by filtering further particles into a web of saliva from a water current generated by wriggling their bodies in the tubes, or by scavenging around at the entrance to their tubes.

Not all midge larvae are red. But those that do contain the red pigment hemoglobin are probably more capable of withstanding conditions in which oxygen is in limited supply. This, together with the fact that the adults of different species emerge at different times and that two or three generations are often produced each year, accounts for the presence of midge larvae in the lake bottom throughout the year.

Related to the midges are the hover flies and drone flies. These have a life history similar to that of the midges but the larvae look very different, and are aptly named rattailed maggots. At its rear end each larva has a pipe or tube, which reaches to the surface of the water at right angles to the body and is used for getting a supply of oxygen. The larvae feed by scavenging around in the mud near the lake margin, where they live, ingesting any suitable-sized particles they find. These may, of course, include microorganisms in addition to the decomposing matter. As they move around, these larvae mechanically break up the leaf litter, so creating smaller pieces on which other animals can feed.

We have now looked at three groups of insects that get their food from the lake floor. There are many others living apparently similar lives, but, as we shall see, they are mainly vegetarians, feeding on living plants, or carnivores, rather than saprophytes. The diet of some aquatic insects is still in doubt, and these groups, such as the caddis fly larvae, will be considered later.

So far we have watched leaves sink into our lake, and we have seen them broken down, eaten, and turned into new living material by a variety of microorganisms as well as some larger creatures. We have also noted that the rotting mass of organic debris and mud provides a home for a rich fauna. However, not all the products of decay

A "bloodworm" (real size under one inch) owes its red color to oxygen-absorbing hemoglobin, which may aid this scavenger's survival in oxygen-deficient mud. The larva pupates, and then becomes a winged adult feather midge.

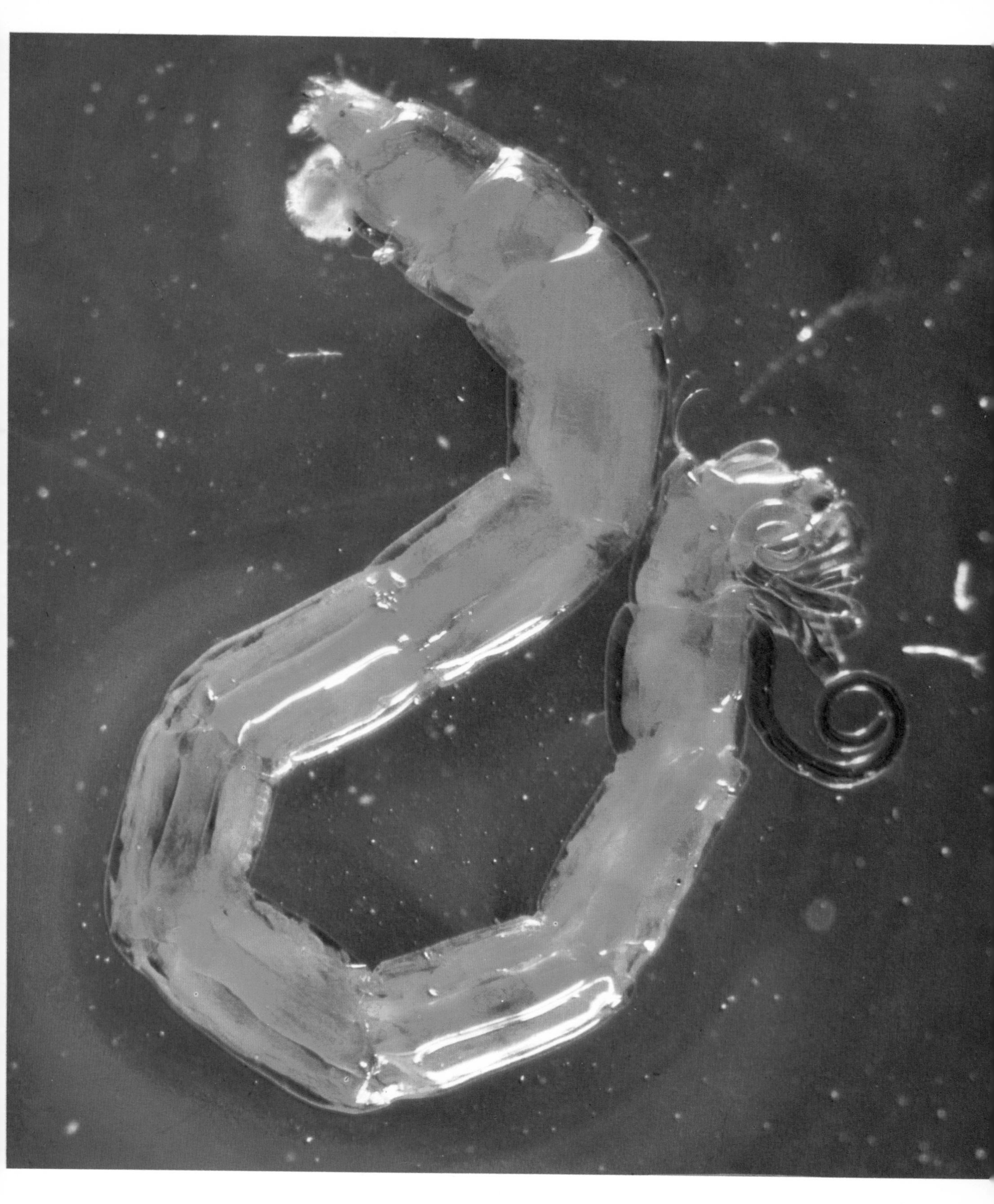

become parts of organisms immediately. Some of the nutrients are released into solution and act like fertilizer in a field or garden. They support the lush growth of many plants—the greenery living in fresh water.

Plants can conveniently be divided, for our purposes, into two groups—the algae and the higher plants, including the mosses. But all green plants have one essential characteristic: they can make living tissue from inorganic molecules, the most basic building blocks for life. This means that from simple, naturally occurring compounds they can make new tissue without consuming anything that has been made by other plants or animals. Green plants are therefore "producers" rather than "consumers." Having made this unifying point, let us have a look at a very few of the thousands of different plants in fresh water.

The algae are a large and diverse assemblage. So much so that some of the groups included may be only distantly related to one another. Indeed, some of the green, flagellate algae have been claimed by both botanists and zoologists although the latter call them protozoans rather than algae. This argument is not of importance here but it does show that where some of these simple organisms are concerned, the dividing line between plants and animals is very vague. Basically algae are unicellular, like the protozoans, but sometimes the cells are clumped together as masses or filaments. Only in the most advanced forms does one find cells within the clump doing different jobs, as do the cells in a human body.

The blue-green algae are very simple organisms that are common in both running and still water. Often cells are found in clumps or filaments, in which case they may be enclosed in a sheath. Some of the kinds that live in clumps have a jelly coating around the cells, and the masses may look like pieces of jelly stuck to underwater objects or floating at the surface. Not all of these algae are blue-green in color, because the pigments may be obscured. As far as we know they reproduce by cell division only, with no sexual reproduction.

Another rather small group of algae in fresh water is the so-called "red" algae, although they are actually green or brown, because the red color that tints their marine relatives is obscured in the freshwater variety. In appearance, the most common freshwater form looks to the naked eye like a string of beads, so it is not surprising that it is often called "bead-moss." In fact it is

Left: tadpoles browse on algal threads growing on the stems and leaves of underwater plants. Freshwater algae are a kind of aquatic counterpart of the meadow grass that sustains many herbivores that live on land. Water snails, limpets, water boatmen, and certain fishes with "vacuum cleaner" mouths all make sessile (fixed) algae a large part of their daily diet.

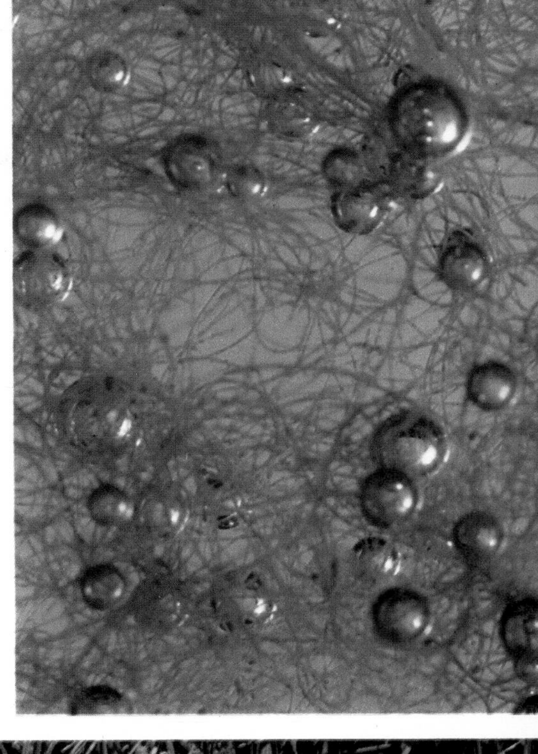

Right: a tangled skein of thread alga. The bubbles are oxygen bubbles, which it gives off as waste in the course of food-producing chemical reactions.

Below: unlike these barnyard ducks, the duckweed and reeds around them can make food from simple compounds in air, and soil or water.

filamentous, with clumps of finer filaments—the beads—at intervals along its length. The alga reproduces by releasing spores from spherical bodies situated on the fine filaments; when the spores burst, a new filament is produced.

It is to two other groups, however—the diatoms and the green algae—that the greater number of freshwater algae belong. The diatoms are unique in having beautifully sculptured shells made of silica, within which they live. They are usually brown or yellow, and in lakes they most commonly form part of the plankton—the floating microscopic life near the surface. Some live on the bottom, however, in long chains or attached to stones or larger plants. As they occur in such different localities, you might expect some notable differences in their physical makeup. Surprisingly, though, all diatoms have the same kind of shell. It is composed of two halves that fit together like a box and its lid. Despite the protective nature of this shell, diatoms are a most important item in the diet of many small animals, especially those animals that live in the plankton.

The largest and most prominent algae of all are the green algae. In this group, the chlorophyll is not masked by any other pigment and so the organisms actually look green. They vary enormously in shape and size, however. Many are unicellular and in isolation are invisible to the naked eye; others form long filaments (blanket weed or thread alga); some live in large clumps; and still others adopt the form of a hollow ball, with the cells arranged around the outside.

Perhaps the most typical unicellular forms are those, such as *Chlamydomonas* or *Euglena*, that some biologists prefer to include among the protozoans. Both of these have flagella—long whiplike threads that are beaten to move them along—but they also contain chlorophyll, just as plants do. Many algae consist of similar units, held together in one of several ways. Thus, *Volvox*, a spherical type, is built of many hundreds, or even thousands, of such units stuck together so that they form the edge of a globe encased in a jelly coat.

The filamentous forms are especially numerous. There are many small differences between each type, of course, but in general they all look like masses of tangled green thread. However, close examination reveals that some of the filaments are straight, whereas others are branched, some grow only from one point in the chain, and others at random from any point. There are also many different forms of reproduction, which permits the different types to be separated. But such differences can usually be seen only with the aid of a microscope.

In general, algae reproduce by cell division, the cells simply splitting in half, but there are also many kinds of sexual reproduction. Often, free-swimming cells are produced by a clump or thread, and those from different clumps fuse together to form a zygote, from which a new plant will develop. The production of a zygote sometimes enables an alga to survive periods of drought, because it can remain in this resistant form for an indefinite length of time.

We cannot leave the algae without mentioning the desmids, which are among the most beautiful of the green algae. They resemble the diatoms in some respects, being about the same size and having the same variety of form as the diatoms. But they are distinguishable by their color—a beautiful green. Another peculiar characteristic is that the cells look as if they are always dividing, as they have a marked constriction around their middle. Many kinds of desmids are to be found in a lake, either living in the plankton or crawling around on solid objects. All desmids are unicellular and can move freely. In bright sunlit water they may accumulate in vast numbers, and the water appears to be an intense green color.

This rather brief look at the algae does not reflect their true importance to the freshwater community, but rather their small size. Their stupendous powers of reproduction permit them to act as something of a safety valve for the community; they can mop up huge quantities of nutrients by breeding very quickly if a surplus becomes available. Larger plants require much longer to breed and do not have this ability to a similar extent. Any sudden increase in algal numbers can be an asset to the animal population by producing large amounts of oxygen as a by-product of photosynthesis, but it may be a menace if the algae then die and a large amount of rotting material is left in the lake, causing a rapid decrease in the oxygen level. As we shall see, the animal plankton depends on free-swimming algae for its food. First, however, let us look at a few of the larger plants to be found in standing water.

Apart from the algae, only a limited number of flowerless plants—the liverworts, mosses, and ferns—are regularly found in permanent standing water, although many frequent the marshy

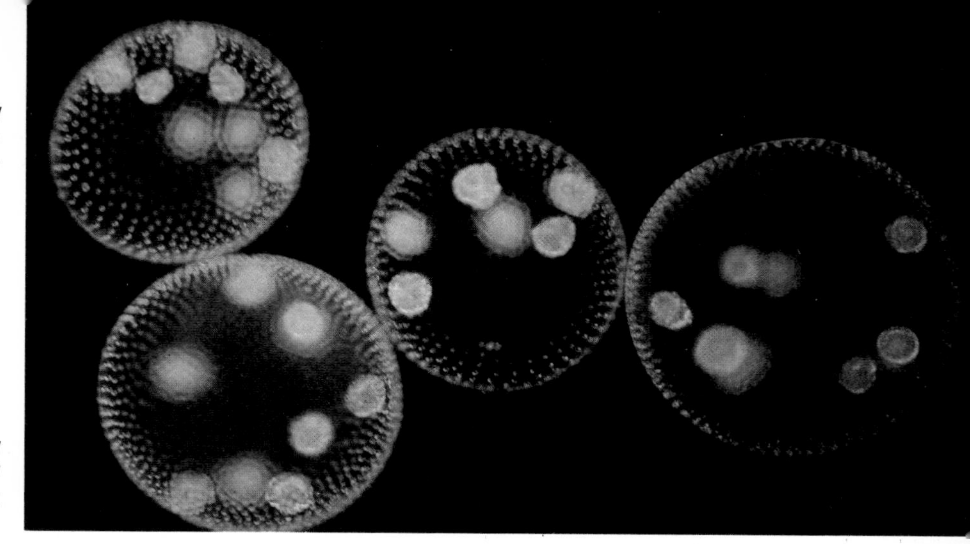

Right: each Volvox *sphere (actual diameter about one tenth of an inch) is a colony of many units whose thrashing whiplike flagella keep it swimming.*

Below: red scum at the edge of this pond is a "bloom" of microscopic one-celled Euglena *algae.* Euglena *and* Volvox *both behave as part plant, part animal.*

edges. Among those found actually in the water is the little floating water fern *Azolla*, sometimes known as "fairy moss." As its popular name suggests, it looks somewhat like a moss, the whole plant being only about half an inch in width, and with each frond divided into two lobes—one floating and the other submerged. But it is the mosses that are most frequently associated with fresh water. The bog moss, *Sphagnum*, is a particularly common inhabitant of the marshy edges of ponds and streams. Forming a thick spongy carpet as a result of its amazing capacity for absorbing water, it is renowned for its bog-building properties.

Equally common is *Fontinalis*, a moss that lives totally submerged in water. This plant is found in both running and standing water and may be anchored or floating, depending on the depth of water, current speed, and similar factors. It has a long, supple stem that may reach over a foot in length and bears many closely packed, pointed leaves. Massive clumps of "willow moss," as this plant is sometimes called, may be found in suitable places, usually where the water is somewhat shaded. Like other mosses, it is able to

How Plants Live in Water

Freshwater flowering plants have become adapted for aquatic life in ingenious ways. For instance, pondweeds have airspaces in the stems and leaves. (In the stem cross section on this page the "empty" areas hold air.) These airspaces help to buoy up the leaves so that they float upon the surface and exchange gases directly with atmospheric air. The floating leaves of spatterdock (the common yellow water lily) "breathe" through tiny pores called stomata, *on the dry upper sides. (The cross section of a spatterdock leaf shows stomata as tiny gaps in the upper surface: large gaps inside the leaf are air spaces giving buoyancy.) Water marigolds exchange some gases with atmospheric air through stomata in leaves that stand above the water surface. But they also exchange gases under water through a different, finely branching, leaf.*

Reproduction poses special problems for aquatic plants: most must somehow grow their flowers and fertilize them above the water level. Cattails produce male flowers above female ones on the same stalk, thus enabling pollen to drift down from the male flowers to fertilize the female ones. Ripe cattail seeds fall into the water and float until they reach a place where they can germinate. Tape grass produces underwater male flowers that break free, float to the surface, and drift about. Pollen may thus reach the female flowers growing at the surface on a long stem. After fertilization, the long stem sinks and seeds develop under water. Tape grass also multiplies by means of runners. Arrowheads multiply by seed, and also produce fat underground tubers.

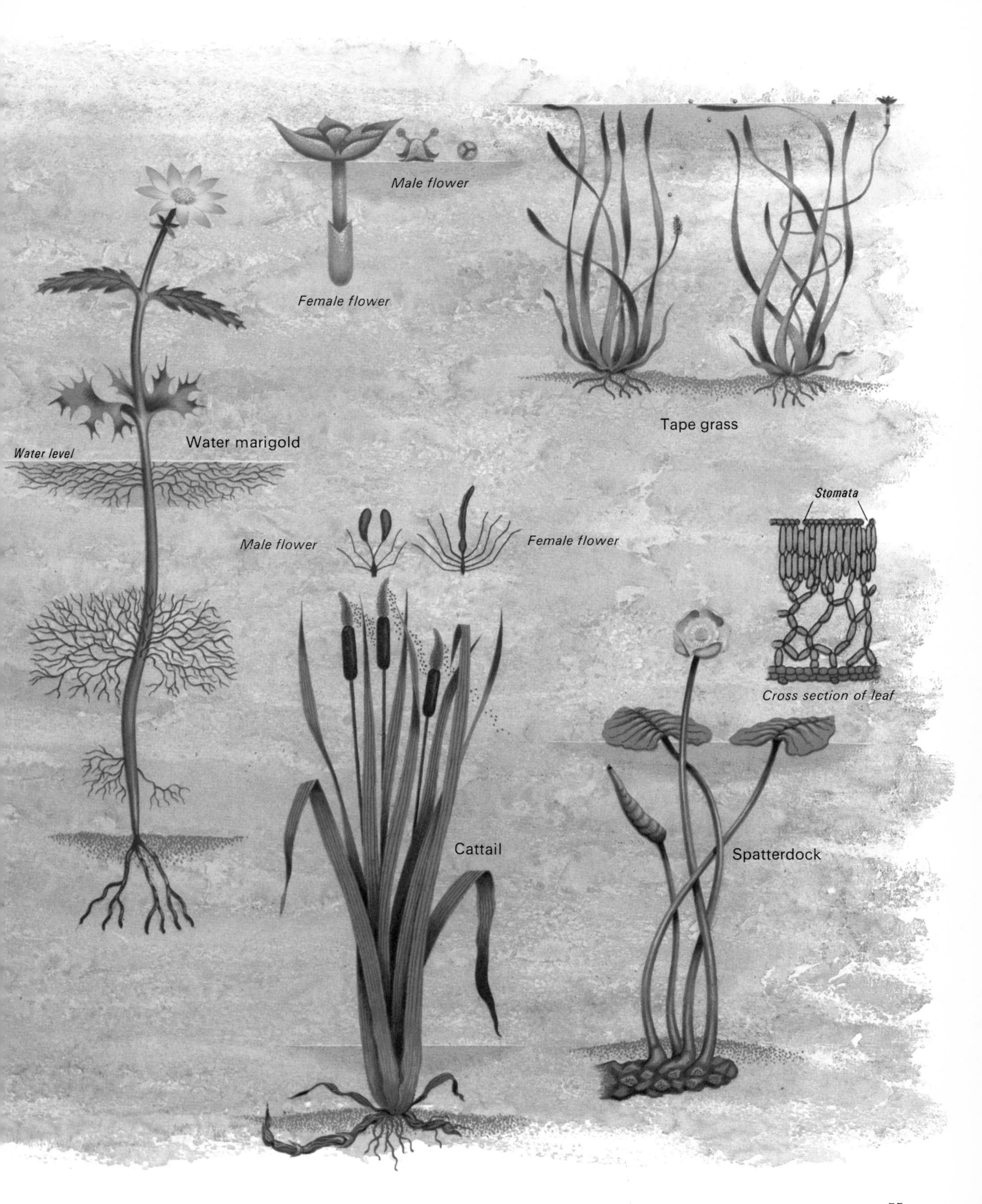
Male flower
Female flower
Tape grass
Water marigold
Water level
Stomata
Male flower
Female flower
Cross section of leaf
Cattail
Spatterdock

reproduce through spores released from capsules borne on the stem, but under normal conditions this method is rare. The vegetative process, in which a piece of plant breaks off and becomes established on its own, is much more common.

Numerous flowering plants are found in lakes, although the flowers are often small and inconspicuous and reproduction may, as in the willow moss, be by the vegetative process.

One of the commonest rooted flowering plants is the pondweed *Elodea*. It is found in standing or slowly moving water, and its branching stems may form very large beds or clumps. The brittle stems with their whorls of short, undivided leaves break easily, and as each fragment is able to regenerate roots and grow into a new plant, the pondweed is able to spread extremely rapidly.

Crowding the water's edge and jostling with the pondweed for growing space is a jungle of greenery. Showy water lilies mingle with water crowfoot, arrowhead, and broad-leaved pondweed *(Potamogeton)*—all rooted in the mud, but raising leaves and flowers above the water surface. Milfoil, recognizable by its long spikes of pink flowers, also competes for space in the water. But it is only the flower spikes that are immediately visible, because the whorls of finely divided leaves are submerged beneath the water's surface. Some of the species of true pondweed, such as the broad-leaved pondweed, have two

Huge "fry-pan" leaves and a flower bud of the giant Amazonian water lily, which blankets quiet Amazonian backwaters.

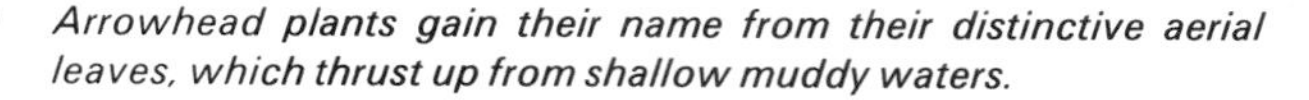

Arrowhead plants gain their name from their distinctive aerial leaves, which thrust up from shallow muddy waters.

types of leaf: underwater straplike leaves and large floating oval leaves. The arrowheads go one step further and have three types of leaf: submerged leaves, floating leaves, and the arrow-shaped aerial leaves from which the plants have taken their name. In its young stages, when the aerial leaves are absent, arrowhead can be easily mistaken for pondweed.

Water lilies are among the most decorative of the freshwater plants. Their long leaf stalks carry the floating leaves to the water surface from the thick stem that remains buried in the mud. Air spaces in the stem are continuous with those in the leaf stalk and the leaves, so that gases can be exchanged throughout the plant. The stem itself grows slowly, and gives rise to new shoots each year. The flowers, usually yellow or white, are surely the most familiar to be seen in water. One species of water lily deserves special mention because of the size of its leaves. This is the Amazonian lily, *Victoria amazonica*, whose circular floating leaves may be up to seven feet in diameter, and will support a small child.

Some of the plants that float on the water's surface are in fact true floaters and do not have roots buried in the mud like the water lilies. Among these the duckweeds are undoubtedly the most common. They include the smallest flowering plant known to mankind: first noticed by a naturalist in the Mato Grosso of South America

The floating leaves of common yellow water lilies sprout anew each spring from tubers in submerged mud.

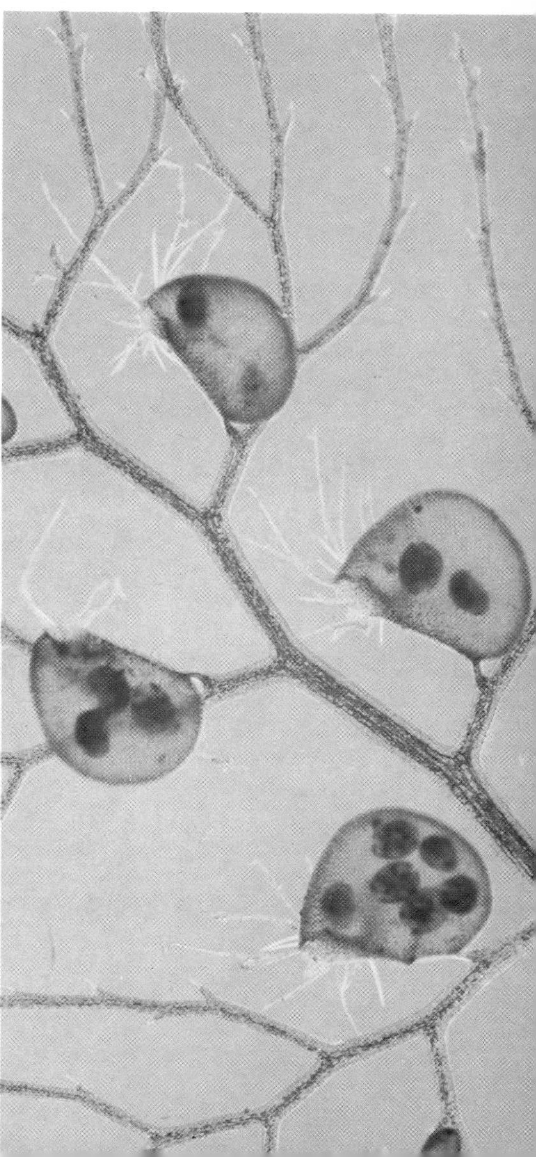

Part of a bladderwort plant, enlarged. The bladders can spring open, trapping and then digesting tiny water creatures.

in the middle of the 19th century, this species of duckweed bears a flower smaller than a pinhead. The "leaves" of duckweeds are not true leaves at all, but pieces of flattened stem. A remarkable feature of the group is their power of reproduction. New plants are budded off from the old, and in suitable habitats they may quickly cover the entire surface of the lake or pond.

Hornworts, too, are common floaters, but being entirely submerged, they are not as noticeable as the duckweeds. They are perhaps the most completely adapted of all plants to living in water. Even the flowers grow under water, and have to rely on the water to transport the pollen.

One of the most fascinating floating plants is the bladderwort with its submerged masses of finely divided leaves bearing rows of death-trap bladders. Here, in complete contrast to the norm, is a case of animals becoming food for plants. The little animals that form part of the plankton community have only to brush against the sensitive hairs at the entrance to one of the saclike bladders, and the trap door opens, sucking in the animal in a rush of water. The door then shuts,

A dense mat of floating aquatic plants conceals this hippopotamus pool in Ruwenzori National Park, Uganda. Cattle egrets perch on the hippos' backs and pick off ticks and other parasites found there.

and the animal dies and its decomposed remains are absorbed by hairs on the bladder wall. After about 20 minutes the bladder is ready again.

Two other floating plants have achieved notoriety in recent years, both as a result of man's activity in Africa. One is the water hyacinth, which reproduced so rapidly in the Nile River at the Gebel Aulia Dam in Sudan that enormous heaps on which people could actually walk were formed. The other is an aquatic fern that also underwent a "population explosion" on several lakes, including Lake Kariba in Rhodesia. Not only did the dense mats of fern prevent fishing by clogging nets—if enough water could be seen to lay the nets initially—but they also caused the death of many fish. Because the plants covered the surface of the lake so thickly, oxygen could find its way into the water only very slowly, and so the concentration fell disastrously. These cases are unfortunately typical of many in almost every country where man has tampered with freshwater communities without understanding the full implications of his actions—and thus has upset the balance of life.

The Lake: Plankton to Fishes

Let us turn our minds back to our lake: we have delved through the mud and seen how dead leaves are broken up. We have watched the nutrients contained in those leaves being converted back into living tissue by a variety of microorganisms, small animals, algae, and higher plants. These creatures have utilized the apparent waste; and the cycle of freshwater life has turned a few degrees. These are crucial stages, however. As we shall see, everything else depends on the organisms that we have mentioned so far. We find one of the most conspicuous examples of this dependence in the world of plankton—a world at which we shall now take a look.

Plankton consists of many tiny plants and animals, known as *phytoplankton* and *zooplankton* respectively, which appear to drift about in the surface waters at the mercy of the wind and currents. A great many of these organisms, however, do have some means of moving independently, and regular changes in the depth at which we find most plankton reflect this ability. The animals living in the plankton depend partly on one another for food, but all of them rely ultimately upon the phytoplankton, which consists of many algae similar to those that we have already briefly considered. Although we shall now concentrate on the zooplankton, we must not forget that the phytoplankton is much more diverse than is often imagined.

Protozoans of many kinds form a part of the zooplankton. Some are rhizopods, such as the sun animalcules we mentioned earlier. But ciliates are the most numerous group, other than the flagellates (which can be considered as plants, and hence part of the phytoplankton). Having already considered their major features, we shall not dwell on the protozoans but look at some larger animals that can just be seen with the naked eye.

Among the most common of the larger animals in plankton are the rotifers, which are in general the smallest of the many-celled, or *metazoan*, animals. This group exhibits a great range of

Sunlight burnishes the deceptively beautiful surface of Lake Erie, one of the world's worst-polluted lakes. But even here, the cycle of life persists in a severely restricted fashion.

The Variety of Habitats in a Pond

Floaters (plankton)
1 Mosquito larvae
2 Volvox
3 Heliozoan
4 Water flea
5 Cyclops
6 Diatoms

Swimmers
7 Backswimmer
8 Perch
9 Pike
10 Great diving beetle
11 Water boatman
12 Carp
13 Brown bullhead

Stone- and weed-dwellers
14 Moth larva
15 Flatworm
16 Water beetle on filamentous alga
17 Hydra
18 Great pond snail
19 Leech
20 Dragonfly larva

Bottom-dwellers
21 Water scorpion
22 Bloodworms
23 Fungi, protozoa, and bacteria

24 Willow
25 Sedge
26 Cattail
27 Water lily
28 Pondweed
29 Hornwort
30 Pondweed
31 Water moss
32 Detritus

From its surface waters through the middle depths to the bottom layers, a pond affords a home to many differently adapted kinds of plant and animal. The upper level teems with tiny organisms collectively called plankton. *Some kinds of plankton appear here, much enlarged (1–6). Planktonic animals (and some "plant-animals" such as* Volvox*) can move about, although the movement is often slight and largely random. Many larger pond creatures, however, are powerful swimmers. Such mobile animals include agile insects (7, 10, 11). Among the most purposeful of movers are the fish; different species seek food at different levels in the water (8, 9, 12, 13). Other creatures spend much of their lives clinging to stones and weeds (14–20), or on the bottom (21–22). The illustration also shows that each level of a pond supports distinctive kinds of plant (24–31). Plant and animal remains sink down to the pond floor to form detritus (32), food for microscopic fungi, protozoa, and bacteria (23).*

body forms. Some drift about in the plankton, some are sedentary and build gelatinous tubes into which they can retreat if disturbed, and some build tubes of mud. They are not segmented as worms and insects are, but they do have definite mouths, guts, excretory systems, and muscles. They have no breathing organs, and the exchange of gases takes place through the body wall. Around the "head" are two bands of cilia that, when moving, look like a rotating wheel—hence the name *rotifer*. Cilia are used for both locomotion and feeding. They sweep food into the mouth, where it is chewed up by hard "teeth."

One of the most interesting features of the rotifers is their mode of reproduction. In many species, only females are known. Of the few males that have been studied, some consist of little more than a mobile bag of reproductive structures, but others are more complex, and more or less resemble the females. Generally, there seem to be two different kinds of reproduction associated with different types of female. The first type never mates, but produces eggs that hatch into females without being fertilized. This form of reproduction is called *parthenogenesis*. The second sort of female—which looks rather different from the first and can be found only at certain seasons, usually in spring or autumn—lays eggs that develop into males. The males mate with females like their mothers, which then lay fertilized eggs. These "resting" eggs remain dormant for a while before they hatch into more

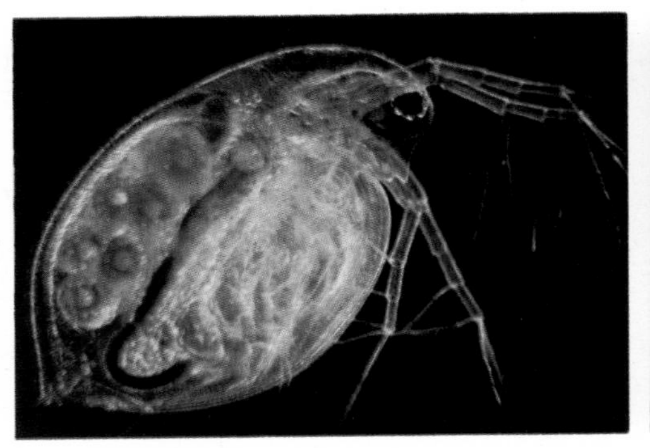

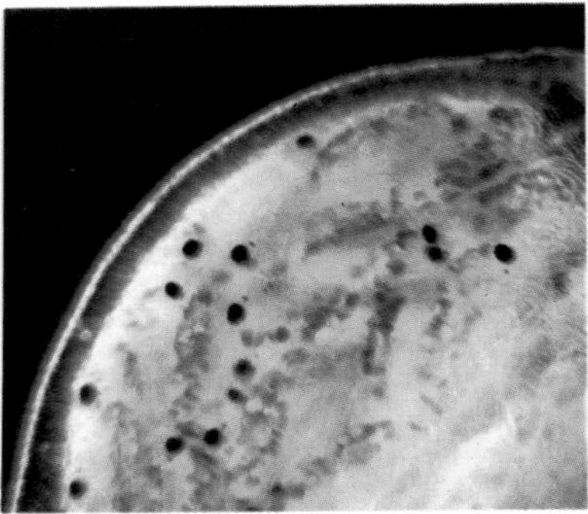

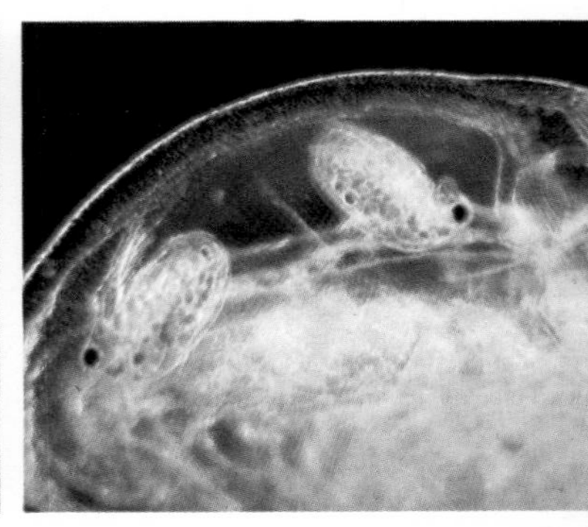

Above: egg development in a water flea's transparent body. Left: early on, eggs appear as shapeless masses inside the mother. Center: eggs have developed into embryos with eyes. Right: the water-flea embryos fully formed.

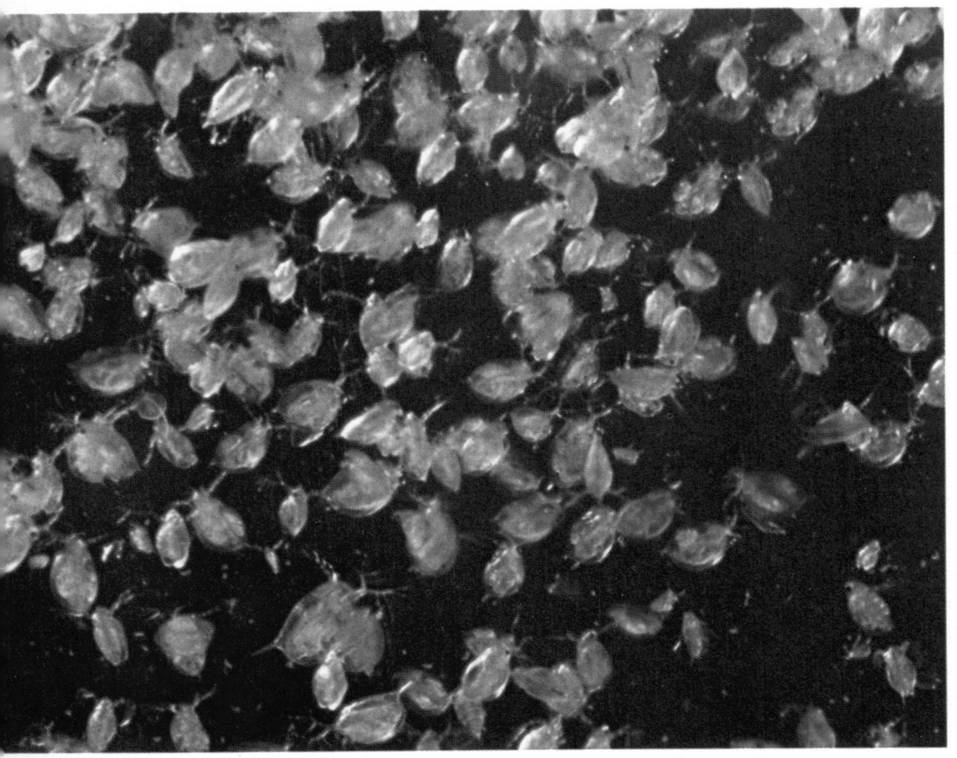

Left: part of a dense swarm of water fleas. These tiny freshwater crustaceans build up into huge numbers under favorable conditions—that is, when their food supply of small green algae is plentiful. The name "water flea" is incidentally misleading. Water fleas superficially resemble true fleas and they move with somewhat flealike jumps. But fleas are actually insects and therefore belong to an entirely different group of animals. Water fleas are more accurately known by their scientific name of Daphnia.

females. If the lake dries up, the resting eggs can survive, and will hatch when water returns.

A second group with many representatives in the plankton of a lake is the crustaceans. In this widely diversified assembly of animals, we have already looked at the freshwater shrimps, which live on the surface of the mud, and the water lice, which live in the mud, but there are at least four kinds of crustaceans to be found in plankton.

Fairy shrimps, for instance, are surface-water dwellers. These animals are relatively large—about an inch in length—and slow-moving. They therefore fall prey to fish very easily and hence do not occur where fish are plentiful, but can often be found in temporary pools, or pools that sometimes freeze solid. They swim on their backs, feeding on algae that they filter from a current of water that passes between their legs toward their mouths. Most fairy shrimps are colorless or pale orange, except for the eyes, which are black, and the gut, which appears green because of the green algae inside. Sometimes their eggs are brightly colored and quite conspicuous.

Far more widespread than the fairy shrimps are the water fleas and their relatives. These animals are common even when fish populations are large, and indeed they represent a most important item in the diet of many fry and small adults. In summer they reproduce so fast that large swarms of them may tint the water orange or red. Each animal has a large, hardened envelope or *carapace*, which is wrapped around the whole body,

Like water fleas, copepods are common pond crustaceans, each no more than a fraction of an inch long. The pear-shaped female Cyclops *shown above bears two relatively huge bags in which egg masses can be seen.* Cyclops *has a single eye between its feelers, and takes its name from the cyclopes—one-eyed giants who figured in ancient Greek mythology. (*Cyclops *is the Greek for "round-eye").* Diaptomus, *a closely related creature (right), has two vastly elongated feelers that help to jerk it through the water.*

Fairy shrimps are inch-long crustaceans that swim slowly in an upside-down position. Their soft bodies and sluggish progress make them easy prey for larger animals. Their leaflike forefeet are used for filtering out particles of food.

and is shed periodically when the animal molts. Two prominent branched antennae emerge from the part of the shell that protects the head. These are used for swimming, which is characteristically jerky, because the antennae alternately work, propelling the animal upward and forward, and rest, causing it to sink back. All the water fleas, of which there are many kinds, filter their food from the water, but (unlike the fairy shrimps) different limbs do different jobs in the filtering process. Two pairs of limbs form a pump or filtering apparatus like that of a fairy shrimp, but the system is closed behind by a further pair, and two other pairs prevent large particles from entering the mouth. Their diet consists almost entirely of small green algae.

Like the rotifers, water fleas produce two types of eggs. The "summer" eggs—that is, eggs laid when breeding conditions are most favorable—contain little yolk and develop rapidly in the mother's brood pouch without fertilization, but "winter" eggs contain much yolk, develop slowly, and require fertilization. These "winter" eggs are fertilized in the brood pouch, and the walls of the pouch then thicken to form a protective capsule around the eggs; this is cast off with the carapace at the next molt. The eggs sink to the bottom, where, after a short period of development, they rest for a while. They can be subjected to drought or ice during this rest period without coming to harm. Although called "winter" eggs, this type may be produced at any time if the animals are subjected to adverse conditions, such as starvation, which could eliminate the entire population. Many kinds of water fleas do, in fact, produce "winter" eggs at two seasons of the year—late autumn and early summer.

One other major group of freshwater planktonic crustaceans are the copepods. There are a great many species of copepods in both sea and fresh water, but all of them lack both the shell or carapace and the large "compound" eyes that are characteristic of most other crustaceans. They occur in a variety of colors—red, green, or blue—though some are almost colorless. One of the most common types is *Cyclops*, which has a pear-shaped body with a rather elongated "tail." At the blunt end of the pear, the head, there is a single, centrally positioned red or black eye and two pairs of feelers, one of which is used by the male to hold the female during mating. The copepods can swim in two ways—slowly by using their feelers, or more quickly by using their legs as paddles. Some, such as *Cyclops*, seize tiny prey with their jaws, but others are filter feeders. Their filtering mechanism is very different from that of the water fleas: they create a feeding current by paddling their feelers and some of their mouthparts, and a fringe of bristles at the mouth catches small particles.

There is no parthenogenesis among the copepods. Eggs must be fertilized by males, after which they are generally retained in sacs beneath the female's body. In most species of copepod these sacs are one of the most prominent features of a "pregnant" female. When hatched, the young do not resemble their parents, and lack a tail, which develops at a later stage. They take about a month to reach maturity. As in the other groups of planktonic crustaceans, such as the water fleas, copepods may lay resting eggs—eggs that can survive adverse conditions by taking an unusually long time to develop.

The fourth group of crustaceans that may sometimes be found in plankton are the ostracods. But they are rarely as numerous as the copepods or the water fleas. At first sight, an ostracod looks like a bean-shaped shell, hinged along its longer convex side, and usually brown or yellow in color. But when the animal is moving about in open water, the shell opens, and the ostracod may paddle along by means of a long pair of feelers. Ostracods can also walk using a pair of legs that can be protruded from the shell. If an individual is disturbed while swimming, it can withdraw its feelers and legs, close its shell, and sink to the bottom. Ostracods may feed on algae or small floating particles, or on pieces of decaying material from the lake bottom. They sometimes reproduce parthenogenetically, just as rotifers do, and the eggs are resistant to drought.

Also often encountered in plankton are the larvae and pupae of flies such as the *Dixa* and phantom midges, and the mosquitoes. These are more mobile and larger than the other animals of the plankton. The most common of them are the larvae and pupae of mosquitoes. The larvae may be up to half an inch in length and are almost colorless. They hang by a breathing tube from the water surface, with their heads downward. It is easy to distinguish the group of mosquitoes to which those that can transmit malaria belong, because the position of the larvae at the surface is different from that of other mosquitoes. They hang down quite steeply, whereas most mosquito larvae rest almost parallel with the water surface. The heads of both types of larvae are equipped with mobile "hairs," which beat rapidly to create a water current from which food is strained. Usually the food consists of algae. When disturbed, the larvae swim downward from the surface, but they soon float up and resume their original position.

When the larva has grown to its maximum size, it metamorphoses—changes very markedly in shape—and looks like a spherical blob with a short tail. This is the pupa, in which the adult fly is formed. The pupa, which does not feed, floats at the water surface where it obtains air through two inverted cones, or trumpets, on its head. When disturbed, it may swim down into the water by thrashing its tail. Eventually, the pupa settles at the surface and stretches its tail; the skin breaks, and the mosquito steps out on to the water before flying off.

Superficially similar to mosquito larvae are the larvae of the phantom midges. However, the midge larvae can be distinguished by the presence of two pairs of air sacs, one pair at each end of the body. These air sacs are hydrostatic organs: they allow the larvae to adjust their density so that they neither float upward nor sink to the bottom. The larva can hover in mid-water without moving and wait for their small prey, which they catch with their antennae. The pupa is rather similar to that of a mosquito, but has a longer tail and narrower breathing trumpets.

One other group of flies whose aquatic larvae are found in the surface waters of lakes are the soldier flies. The larvae of these flies are tapered toward their tails and look rather like brown maggots that have been stretched. Hanging from the surface film by means of a tuft of bristles on their hind-ends, they use their mouthparts for catching the small creatures on which they feed. Often the bristles hold a bubble of air, which allows the submerged animal to remain in the water for rather longer than would otherwise be possible. The pupa is different from the others we

Above: Culex *mosquito egg raft and (left) larvae and pupae, all shown enlarged. The raft consists of individual banana-shaped eggs bonded together by a sticky substance from the mother's body. The eggs float on water for two or three days, then hatch into bristly "wrigglers" like the five shown here, hanging head down from the water surface. The four rounded objects with short tails curled just below the surface are mosquito pupae. They are in the process of becoming winged adults.*

Above: unlike its relative the fairy shrimp, the tadpole shrimp has a protective shield built into its back.

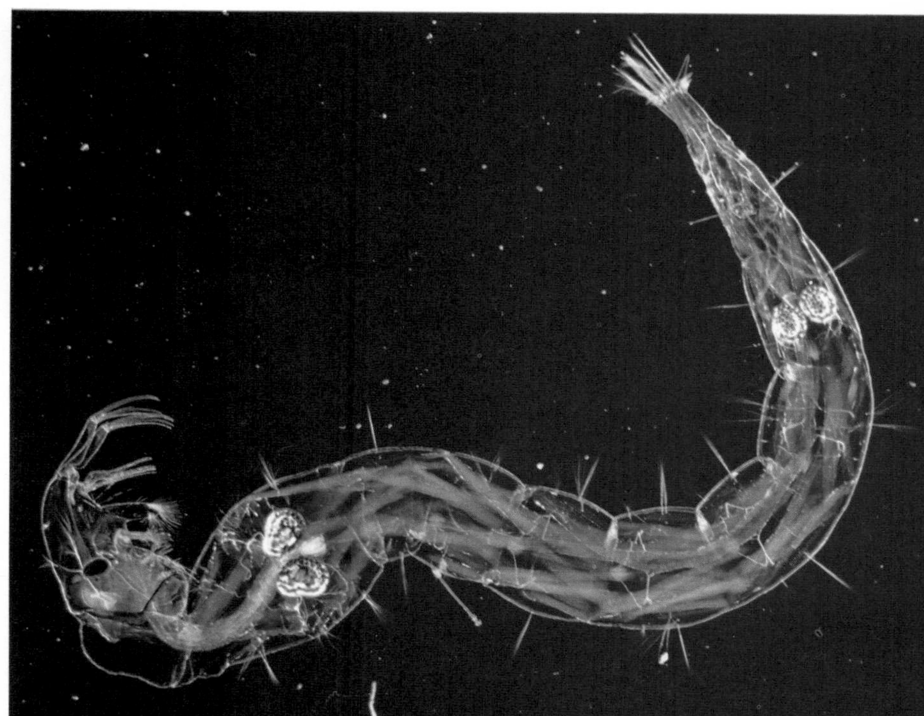

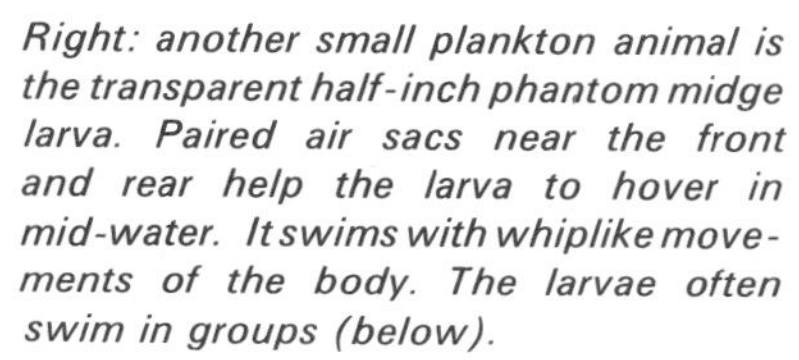

Right: another small plankton animal is the transparent half-inch phantom midge larva. Paired air sacs near the front and rear help the larva to hover in mid-water. It swims with whiplike movements of the body. The larvae often swim in groups (below).

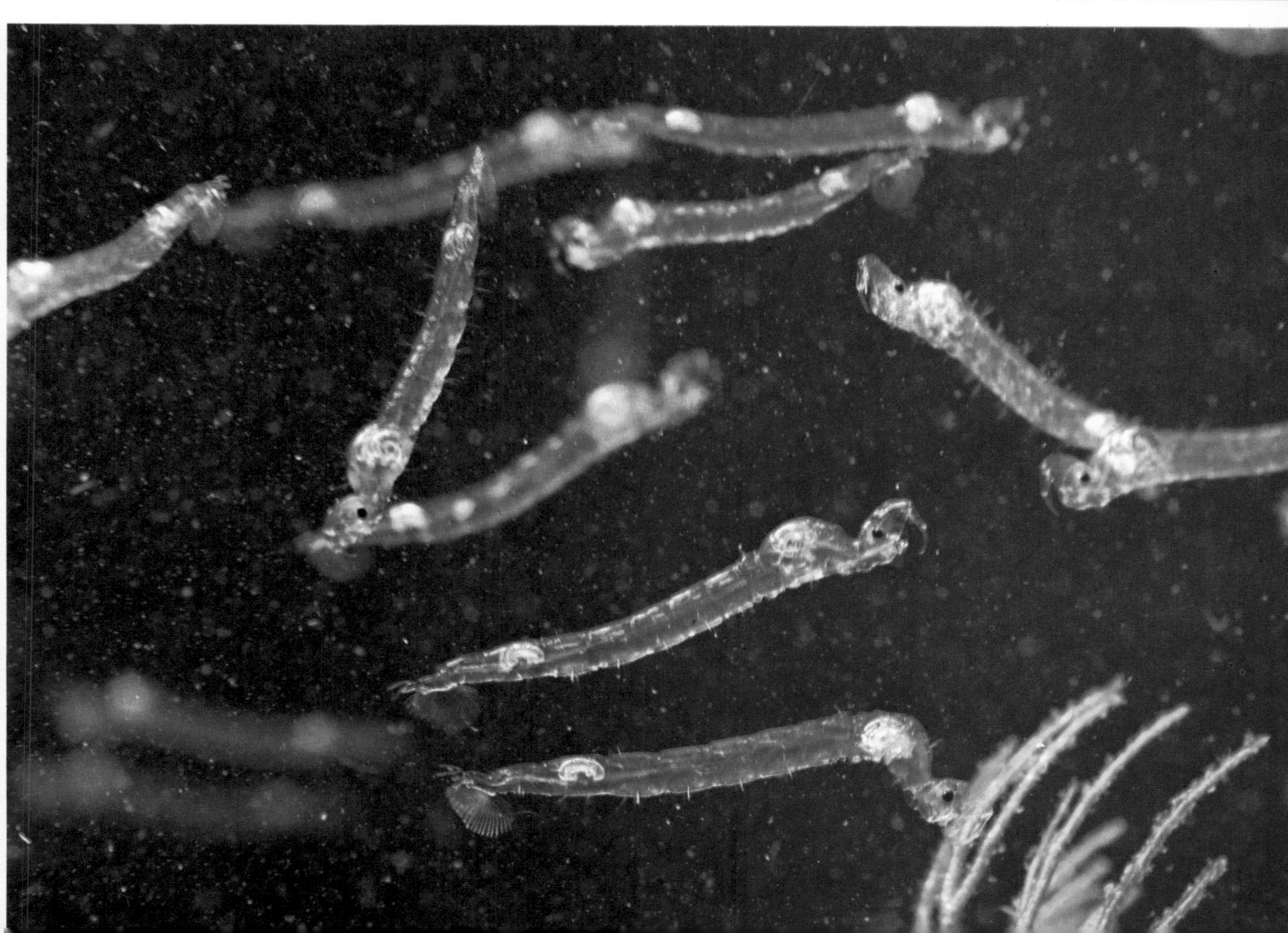

The World Beneath a Lily Pad

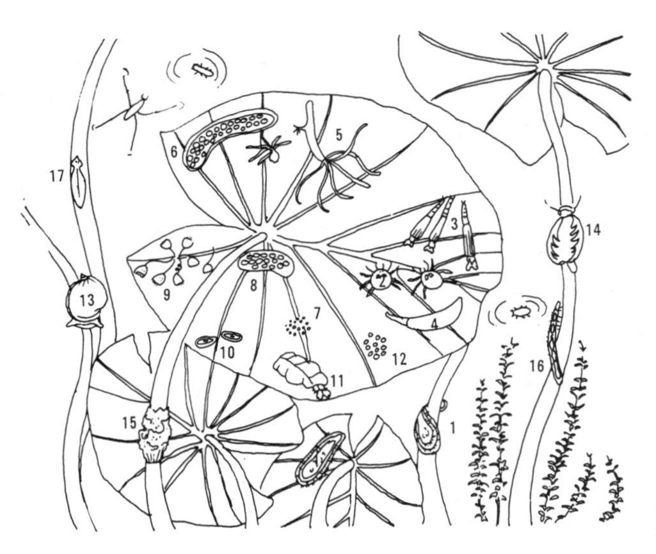

1 Moss animal
2 Water mite
3 Rotifers
4 Beetle eggs
5 Hydra
6 Snail eggs
7 Caddis fly eggs
8 Water mite eggs
9 Bell animalcules
10 Lake limpet
11 Brown china marks moth caterpillar
12 Whirligig beetle eggs
13 Wandering snail
14 Bladder snail
15 Pond sponge
16 Bristle worm
17 Flatworm

A fish's eye view of life beneath a water lily pad shows that even a seemingly bare, flat expanse of leaf and stem can play host to many small members of the pond community (shown here to differing scales). Some creatures spend most of their lives fixed to the surface of a leaf or stem. Such organisms include the moss animals or bryozoans (1), hydras (5), bell animalcules (9) and some rotifers (3), lake limpets (10), and pond sponges (15). Other organisms travel slowly over stems or leaves. Such slow movers include the brown china marks caterpillar (11), wandering snail (13), bladder snail (14), bristle worm (16), and flat-worm (17). More agile are the water mites (2), eight-legged relatives of spiders.

For all such creatures, leaves or stems furnish a supply of food or a foothold on a fruitful hunting ground. For instance, the caterpillar feeds on leaves growing at the water surface. Water snails and limpets eat algae growing on underwater leaves and stems. Freshwater sponges, fixed rotifers, and hydras trap food borne to them by water.

The underside of a water lily pad is also a nursery. Here we may find the eggs not only of such leaf-bound creatures as the water snails and water mites (6, 8) but also of more mobile animals such as caddisflies and water beetles (4, 7, 12).

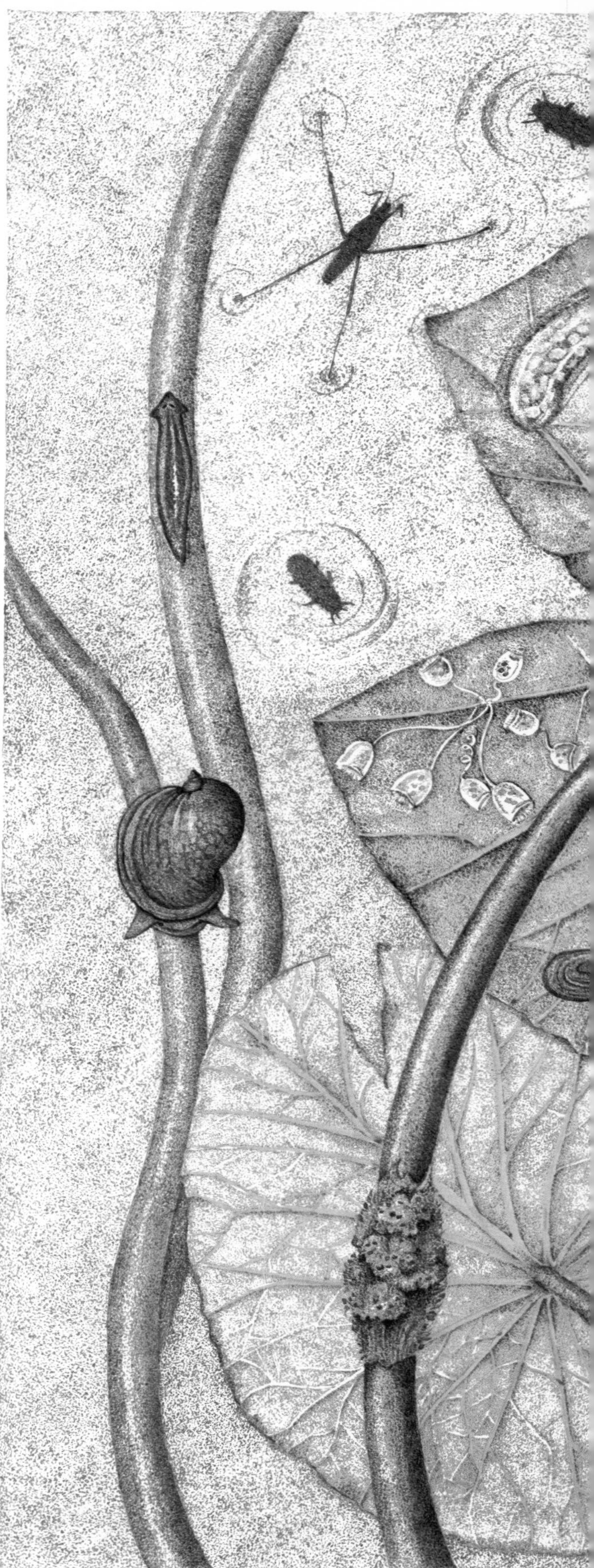

Above: five brown hydras with tentacles extended pose formidable traps for any water fleas or Cyclops *that may jig their way. Barbed or stinging threads hurled from such tentacles subdue the hydra's prey. The tentacles then cram the prey into the hydra's mouth and digestive juices dissolve it.*

Left: the flattened, coiled shell without a spire identifies this as a ramshorn snail. Ramshorns slide over smooth plant surfaces on a muscular foot, rasping algae from plant stems and leaves with rows of teeth on a ribbon-shaped organ called a radula*. These water snails breathe with a kind of lung and must periodically surface to refill the lung with oxygen.*

have considered in that it lives in the larval skin.

The young of larger animals may also be found in plankton, although this is less common among freshwater than marine animals. Many freshwater fish, for example, lay their eggs on or near the bottom, or stick them to plants. Relatively few eggs therefore float with the plankton, where they would have less chance of survival in running water. Other fish, such as the cichlids, sticklebacks, and bass, tend their young and keep them together until they can swim well enough to escape parental control. And so, although we may find some young fish feeding in the plankton during the spring and summer, they are able to move into deeper water if necessary.

The small fish that do utilize the zooplankton for food represent yet another stage in our cycle. We know that the algae are producers of new living material, and that the water fleas are *primary consumers*—they eat the producers; now we have reached the level of the *secondary consumers,* which eat the primary consumers. But before we follow this apparently simple cycle any further we must move sideways to look at the algae-eating animals that live deeper in the water. A simple way of finding these animals is to examine the larger aquatic plants.

The surface of a plant leaf, like a stone or a submerged twig, forms a suitable habitat for the kinds of algae, such as many of the clumped and filamentous varieties, that need to be anchored. Like their planktonic relatives, these algae provide a rich supply of food for some animals. For instance, the snails and their relatives the limpets feed principally upon this algal coating. Both of these animals are equipped with a *radula*—a long ribbon with transverse rows of teeth on it—which is used to rasp the algae away from their anchorage. The radula slowly moves forward as those teeth in use are worn away.

Freshwater snails can conveniently be divided into two main groups. There are some called the operculates, which can seal the entrance to their shells with a plug, or *operculum,* and have gills that equip them for living in oxygen-rich running water. Then there are others that have neither operculum nor gills, but have instead a type of lung that must be filled with air by visiting the water surface. This second group, known as the pulmonates, are the more common in standing water, such as lakes. From their method of breathing it is almost certain that they have descended from land snails that returned to the water. Freshwater limpets are also pulmonates, but unlike the majority of pulmonate snails they do not need to return to the surface for oxygen, and can extract all they need from the water.

All pulmonates are *hermaphrodite*—that is, each animal is both male and female. Most of their eggs are nevertheless fertilized by a second individual before being laid in a tough jelly coating that the animal attaches to a plant, stone, or piece of debris. This much they have in common, but most pulmonates occurring in lakes fall into two distinct subgroups: the true pond snails and the ramshorn snails. The true pond snails, which are usually brown in color, have coiled, spiral shells rising to a point or spire. Of the many species, one—the dwarf pond snail—is of particular importance because it harbors one stage in the life cycle of a sheep parasite, the liver fluke. This species is not normally found in lakes, but inhabits temporary ponds on grassland.

The ramshorn snails can also be distinguished by their shells, which are flattened coils lacking a spire. There are many species of these snails, too, the majority being reddish brown in color. Like the *Tubifex* worms and midge larvae, their blood contains hemoglobin, which enables them to utilize oxygen more efficiently. Some of the tropical ramshorn snails have a justifiably bad reputation among men, for they harbor organisms that cause the disease bilharzia. Both types of snails lay their eggs in spring and early summer, and the eggs hatch directly into miniature adults.

Freshwater limpets, like their marine relatives, have hood-shaped shells that are neither coiled nor spiraled. The shell, which differs from that of seashore species in being much thinner and often slightly hooked at the apex, is held down tightly to flat surfaces by means of the suckerlike foot so that the animal is protected against adverse conditions. Limpets are rather less mobile than snails, but nevertheless their life style is very similar to that of the snails.

Aquatic plants may be home not only to numerous snails and limpets, but also to a number of other animals, which, although they feed neither on the plant itself nor on its algal coating, benefit from the protection offered by the plant, and find it a convenient situation for hunting. To this group belongs *Hydra*—a freshwater relative of the marine jellyfish and corals, all of which are members of the group known as the *coelenterates. Hydra* is a solitary animal consisting of a stalk or body that is anchored at its lower end, often to a submerged plant leaf or stem. The body is hollow, with a single opening at the free end, which is surrounded by a ring of arms or tentacles. Both body and tentacles can be contracted; but even when fully extended, the animal is rarely more than half an inch long. *Hydra* has no special breathing organs, and gaseous exchange takes place through the body surface.

There are many species of *Hydra*, varying in color from brown to green—the green color being due to tiny algae that may live inside the cells of the animal's body. It is the presence of this alga that enables *Hydra* to pass unnoticed on green leaves and stems while it waves its tentacles about in the search for small animals on which to feed. The tentacles contain special cells that can shoot out barbed or poisonous threads very similar to those used by jellyfish to sting, and these threads tie up or impale the *Hydra*'s prey. Once trapped, the food—often a water flea or copepod—is wrapped up in the tentacles, which stuff it into the mouth. Digestive juices are released into the body cavity, the food is "dissolved" and absorbed, and the waste ejected through the mouth.

Hydra can reproduce in two ways: young animals may grow or "bud" from the side of an older animal and then separate away at the base to start a life of their own; or sexual reproduction may occur. In the sexual process, sperm, produced from a bulge on the stalk, swim up to the eggs, which lie in another bulge nearer the mouth. Once fertilized, the eggs are encased in a firm "shell" and either sink to the bottom or stick to plants. Most sexual reproduction seems to occur in autumn. Occasionally, although rarely under natural conditions, complete new animals can grow from small pieces of one specimen.

Hydra catches its prey among its tentacles and generally stays in one place for considerable periods, waiting for its meals to come to it, whereas the flatworms, which share the same sort of habitat, are more active animals in that they move about in a constant search for food. These animals, about half an inch in length, are among the most numerous freshwater organisms. Their jellylike bodies, varying in color from black through brown and red to almost pure white, can be found beneath nearly every stone, and crawling on most plants. They move by creeping along on waves of cilia without distorting their bodies unless in a real hurry. This means of locomotion is enough to indicate that they are not related to the true worms that we find in mud. Their nearest relatives are the parasitic flukes and tapeworms.

Flatworms feed on any dead or living animal matter that they can obtain. They have no mouth at the front of their bodies but protrude a tube from their underside with which they can suck up pieces of food. As this tube is the only opening to the gut, solid waste is excreted through it between meals. Flatworms find no problem in obtaining oxygen, as gas exchange takes place over the entire surface. Like *Hydra* and the pulmonate snails, they are hermaphrodite, but a second individual always fertilizes the eggs before they are laid. The eggs themselves are laid in groups in cocoons attached to suitable underwater objects. Flatworms can replace parts of their bodies that have been lost or damaged. Thus, if a flatworm is cut in half, the head will grow a new tail and the tail a new head.

We can expect to find some snails, coelenterates, and flatworms in almost every lake. But other creatures are less cosmopolitan. This does not mean that they are less well adapted for life in fresh water, but rather that they represent unusual invasions of fresh water.

In Europe, for instance, there lives a spider that spends its entire life in fresh water. Many other spiders live in damp margins around water, but the water spider is the only truly aquatic species. In appearance this spider is very similar to its terrestrial relatives, but its body has a silvery sheen because of the air that is trapped around it, held by tiny bristles. The male is just over half an inch long, and the female rather smaller. Each sex builds a web of silk, incorporating a few pieces of plant material. Initially, the web is slung horizontally between upright supports—plant stems, for example—but later the spider collects air from the water surface, carrying it between its legs and abdomen, and releases it beneath the center of its web. The effect of this procedure is to carry the center of the web upward until an air-filled, bell-shaped structure is formed. The spider then lives in its bell, breathing by means of the air supply inside and emerging only to collect dead or living animals, which are carried back to the shelter before being eaten. Mating takes place in summer, and the female subsequently lays her eggs inside her bell. The newly hatched spiders often live in empty shells before building their own bells.

This fascinating spider is well worth studying if you have a chance. A specimen can easily be kept in a jar, where it will build a web and live an apparently normal life if supplied with such suitable food as small insects. Unlike many other freshwater animals, these spiders can be found in the winter; they survive the cold season in a bell built deep in the water, where it is not affected by ice. When the water warms up again, the animal becomes active.

Another group of inhabitants of some larger water plants are the caterpillars of a number of different moths, such as the china marks moth. It is perhaps surprising to find that the familiar moths of garden and hedgerow have aquatic relatives, but there are a number to be found in suitable localities. Different species tend to have their own specific host-plants but most of them have a similar life history. The moth lays its eggs on a leaf at the water surface, and from these eggs the caterpillars hatch. The young caterpillar builds a home from one or two pieces of leaf tied into place with silk on the underside of a floating leaf. It lives in this leafy tube, feeding on more leaf as it grows.

The newly hatched caterpillar possesses no breathing organs, and absorbs as much oxygen as it needs through its skin from the surrounding water. But as the caterpillar matures, its body becomes covered with tiny bristles so that the skin surface is unwettable, and it cannot absorb oxygen from the water. By this stage, however, the inside of the caterpillar's tubular case is quite dry, and the larva is able to breathe the air that surrounds it inside the case. In this situation a pupa or chrysalis is formed. After some time, the adult moth emerges at the water surface and flies away to mate. The female then returns to the water to lay her eggs.

Now that we have taken a look at some of the animals associated with the larger aquatic plants, an unexpectedly complex picture of the pond or lake is beginning to emerge. We have seen that algae and higher plants depend on nutrients from the mud and that floating algae—the phytoplankton—support a rich fauna—the zooplankton. Moreover, in getting a broad view of the many animals that live in association with larger plants, we have become aware of another important point: that no possible food source or microhabitat remains unexploited. We have seen spiders collecting food, *Hydra* using plants as a platform on which to live while grabbing food coming past, and flatworms tidying up the plants by eating animal matter off the plants' surfaces. Even the thin algal slime supports an assortment of creatures such as snails and limpets.

All these are examples of animals utilizing the apparent loose ends of nature's web of life. They are by no means the only examples. But now it is time to move along up the size scale, to the larger animals that are much more familiar to most of us.

Among these larger, more familiar animals living in fresh water are the water beetles. Adults may be up to one inch long, but whatever their size they all share common characteristics: very rigid bodies with obvious heads and strong, biting jaws; and hard upper wings folded over the abdomen, concealing a pair of large membranous wings for flight. The larvae show more variation but usually can be recognized by the prominent head and jaws, clearly segmented body, and legs that are often flattened and fringed with hairs, their paddle shape giving an increased efficiency in swimming. The pupae of water beetles are usually noticed out of water, because mature larvae walk up the bank or along emergent vegetation before metamorphosing into pupae.

There are many species of water beetles, some of which are strictly vegetarian, feeding exclusively on filamentous algae or parts of the larger aquatic plants. If you take a close look as a mass of algae or larger plants, you will find many such beetles in it. Others eat both plants and animals; and still others—a great many, in fact—are voracious carnivores that feed not only on snails, smaller insects, and each other, but even on small fish or tadpoles. The beetles' method of feeding is interesting. They do not always use their strong jaws, which are found in both larvae and adults, for chewing up their prey. Instead, they pour digestive juices into the food through channels in the jaws, which have first pierced the plant or animal. These juices convert the soft, nutritious parts into a thick, soupy mixture, which is then sucked into the gut. Even some of the largest water beetles, the dytiscids, eat in this manner.

Beetles have several different ways of obtaining oxygen for their underwater existence. Some of the larvae have "gills" on their abdomen and can extract oxygen from the water, but the majority of both larvae and adults swim up to the water surface to breathe. By storing air in different parts of the body they reduce the frequency of these visits. Some of the large dytiscids, known as the great diving beetles, store air between their wings and abdomen, and others carry a conspicuous air bubble at the tip of the abdomen. Carrying air in this manner makes the animals buoyant, and so they must constantly hold on to something beneath the surface to avoid floating upward.

Extreme weather conditions do not trouble water beetles. In winter, if the ice cover becomes too severe, both larvae and adults burrow into mud at the lake bottom or around the lake's margin to wait for better weather to return. In summer, because the adults can fly, they are usually among the first animals to colonize a new body of water such as a freshly filled garden pond. Drought is not, therefore, a real hazard to adults, although many larvae, which are less mobile than their parents, may die. In water, however, both larvae and adults can swim efficiently, especially the carnivorous forms that must be able to move quickly to catch their prey.

Although they cannot be considered especially attractive in appearance, beetles do not look ferocious; but the same cannot be said of the water scorpions. These fierce-looking animals are in fact harmless and they are totally unrelated to scorpions. They are simply another type of aquatic insect, closely related to the water boatmen, whereas land scorpions are not insects but

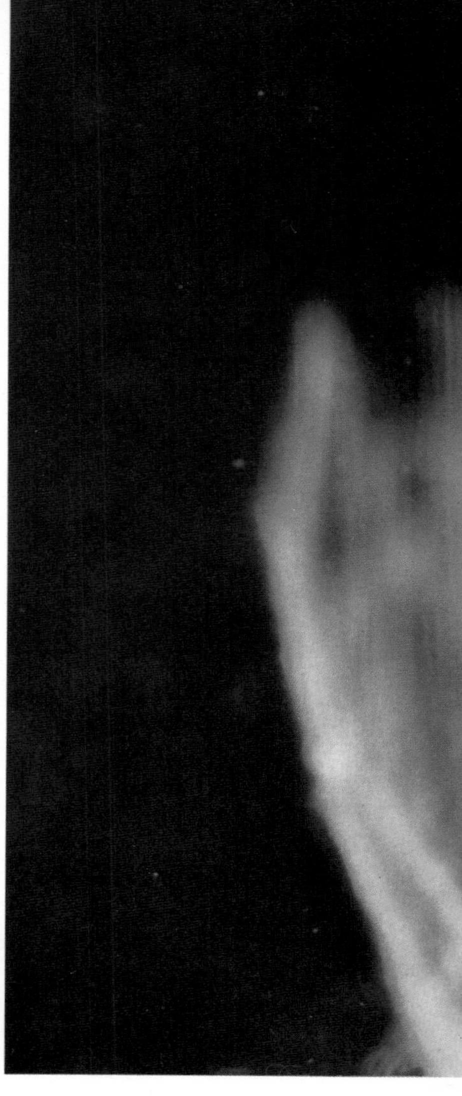

Water spiders and giant water bugs are two of the fiercest predators among freshwater invertebrates. Above: a water spider seizes a water louse that has strayed into its underwater web. Left: a giant water bug descends upon a common newt, driving the sharp claws of its powerful forelimbs deep into its victim's body.

are related to spiders. One group of water scorpions, the only ones found in Europe, are quite small. Some of them are roughly oval in shape and about three quarters of an inch long, and others are about twice as long but very slim. Both kinds have long, straight breathing tubes extending backward and upward from the rear tip of their bodies, and their front legs are used to capture and hold their prey. Superficially, the breathing tubes look like a scorpion's sting, and the front legs like a scorpion's pincers.

Water scorpions are most common around the margins of lakes in shallow water, particularly among vegetation or detritus. They seize their prey with the front legs and suck out the soft internal organs through a sharp proboscis. They do not seem to be fussy feeders and will eat any small animal they can catch. Some water scorpions that live in America are much larger than the groups just described: known as giant water bugs (or, for an unfortunately good reason, as toe biters) they may be more than two and a half inches long. Similar in appearance and habit to their smaller cousins, they are carnivorous but can take a wider range of prey, including snails, fish, and tadpoles.

Close relatives of the water scorpions are the backswimmers. These are familiar animals, about three quarters of an inch long as adults; you can often see them floating at the surface of the water with their undersides uppermost. In immature backswimmers the abdomen has a greenish tint, but it appears silvery in the adults because of a layer of trapped air. The adults are winged, and the head has a pair of prominent brown eyes. Unless anchored under water or actively swimming, backswimmers float to the surface, where they rest with their heads slightly lower than the tip of the abdomen. As with other insects, they are not in danger of drowning, because insects do not

Above: a great water-beetle larva uses its fanglike mouthparts to seize a tadpole. The larva injects juices to liquefy the victim's tissues, and sucks them out.

Above: the upside-down posture of the backswimmer gives this three-quarter-inch water bug its name. Backswimmers row through water with swift strokes of their long flattened hind limbs and use a powerful proboscis to kill prey.

Left: a water scorpion clutching its victim, a damselfly larva. Water scorpions (some exceed two inches) eat any small animal they can catch. They seize it in forelimbs that work like pincers, bite it, and suck out body juices through a sharp proboscis.

breathe through their mouths. Backswimmers are active carnivores, feeding on animals that may on some occasions be larger than themselves. The body contents of their prey are sucked through a sharp, needlelike proboscis, which can also be used to give human beings a painful prick.

The water scorpions and backswimmers have a number of relatives with similar carnivorous habits, but none is as obviously ferocious as the larvae of dragonflies and damselflies. These animals are exactly what they appear to be: savage predators. They are mainly dull brown or green, with large eyes for spying out their victims while they wait concealed among the rotting vegetation on the lake bottom. When the intended meal—a freshwater shrimp, water louse, or bloodworm, for example—is at suitable range, they shoot out a long structure with a pair of hooked jaws at its tip. The hooks grab the prey and draw it into the mouth, where it is chewed and swallowed. This structure covers the lower part of the larva's face when not being used, and it is therefore called a "mask."

Larval dragonflies can readily be distinguished from larval damselflies. The former are short and fat, with very short projections from the rear of the body. They get oxygen by pumping water over "gills" in the gut. Damselfly larvae are generally slimmer and have long projections that act as gills, because oxygen is obtained from the water through their thin walls. When the larvae of both these creatures approach maturity, wings begin to develop above the abdomen. Ultimately—perhaps two years after hatching from an egg—the larva climbs up the stem of a plant until it is above the water and rests. The skin splits over its back, and the adult emerges from the dull larval skin. The contrast in appearance is staggering as the beautiful red, blue, green, or yellow winged insect emerges.

The combined efforts of these killers of the insect world can greatly reduce the population of smaller animals that we have seen slowly being built up from decaying leaves. The predators themselves are controlled by their own greed. If one group eats too many prey, starvation follows, and so the number of predators falls until the amount of food for them increases once again. Conversely, an increase in the available food will permit the population of consumers to rise.

This dragonfly larva devours a tadpole caught by thrusting out an armored "lower lip." Dragonfly larvae can walk or they can swim by using an aquatic form of jet propulsion.

The beetles, bugs, dragonflies, and damselflies are not the only insect predators in lakes. The larvae of alder flies, for instance, are common inhabitants of mud and dense vegetation at the bottom of lakes and slower stretches of rivers. The alder-fly larva enters the water soon after hatching from an egg laid near the water on a plant stem. It has a large head and a pair of powerful jaws. The body tapers away from the head, and the abdomen is bordered by long feathery filaments extending sideways from the body. These are the animal's gills. The young larva actively hunts and feeds on a variety of small animals. Older larvae, nearing pupation, tend to be more sedentary in their habits. The mature larva leaves the water and digs a shallow pit in soil near the lake in which to pupate. The adult, which you can see near water, is a dull creature, usually brown in color; it lacks the grace or power in flight of many other insects, and spends much of its time resting on waterside plants.

In the typical freshwater lake there are also plenty of predatory invertebrates that are not insects. One of the other groups that many people do not consider attractive is the leeches. Contrary to common belief, not all leeches suck the blood of man and other mammals; a great many live on a diet of worms, insect larvae, and snails. Leeches are very closely related to the segmented worms found in the mud on the lake bottom, but they differ from them in having two suckers—one at each end of the body—with which they can attach themselves to stones or plants. Once attached, they look like lumps of jelly. In color, they vary from black through brown, red, and green to yellow or even a creamy white, and they are often patterned with dots or stripes. Sometimes the branched gut can be seen through the skin; it extends back from the mouth, which is in the middle of the sucker on the head.

Leeches swim quite strongly by throwing their bodies into series of curves. More often they "loop" along, alternately attaching the two suckers, extending the body between each attachment of the front sucker, and then bringing the rear end up to it before attaching the other. There are two major and easily recognizable groups of leeches. The first group can be distinguished by the proboscis and the absence of jaws. A common representative of this group is *Glossiphonia*, which, depending on species, may be green, brown, or yellow in color. It sucks blood from snails or midge larvae, as do a number of

Damselflies mating. The male (the upper insect) grasps the female's "neck" with clasping devices at the end of his abdomen. The female binds her own abdomen forward to reach the male organ. This curious union permits mating to occur in flight.

other leeches in the same group. A few are fish parasites or bird parasites. The second group have jaws with which to gnaw into their prey. Horse leeches (so called because they were once thought to attack horses in the act of drinking) and medicinal leeches (formerly used by doctors for bloodletting) fall into this category, but a great number feed on invertebrates. Some suck blood from worms or flatworms, and a few swallow such animals as tiny crustaceans or insect larvae.

Having looked at these predators, let us briefly summarize what we have seen. We began with leaves that decay and are eaten by a wide range of microorganisms. The nutrients are ultimately reconstituted into living tissue, whether plant or animal. Floating algae form the basis of a complex planktonic community. The larger plants shelter a number of carnivorous animals, and snails browse on the algal slime on the plant leaves. In turn, both the plants and the animals, even including the smaller predators, are eaten by larger invertebrates such as beetles, bugs, dragonfly larvae, and leeches. Effectively, then, chemicals derived from plant leaves have been utilized as a source from which quite large animals can arise. We have come a long way already, and many tiny organisms have been seen to play important roles in the lake community; but on this array we can now superimpose the animals with backbones—so-called *vertebrates*, primarily the fish.

A major difference between the fish and the insects and their kin is that the fish have an internal skeleton, which is more flexible than the rigid box or *exoskeleton* that covers the arthropods (such as insects, crustaceans, spiders). So what roles do freshwater fish play? Many people know something about some fish—those that they catch, perhaps, or those that they keep in bowls or tanks. And a few types have found fame for other reasons—the piranha and the sturgeon, for example. But there are many less familiar kinds.

Not all the freshwater fish are "top" predators—that is, predators that lack larger enemies that are able to consume them. As well as those in this fortunate position, there is an enormous

The man and children at this Mexican river may be unaware of the varied and fascinating lives of its many invertebrate inhabitants. For them, as for millions of anglers all over the world, the thrill of fresh water lies in catching fish.

Left: the bowfin is the sole survivor of a fish family some 130 million years old. Bowfins feed on other fish and crayfish and may grow to more than two feet long. They live in sluggish weedy rivers of the Great Lakes and Mississippi systems.

Below: garpikes, like bowfins, are direct descendants of an ancient group of bony fish. Gars once lived in Europe, Asia, and North America. Today, these long, lean, freshwater predators survive only in North America southward as far as Cuba and Mexico.

range of fish that have evolved in such a way as to utilize a great many food and habitat sources. Some feed on invertebrates, some on plant material, some on detritus, others eat fish, and a few depend on food from outside the water.

One important point must be remembered: in talking about "our" lake we do not specify its geographical location, for the simple reason that the same factors and similar animals occur in similar lakes everywhere. The differences between the leech populations of comparable lakes in America, Africa, and Britain are small at the level at which we have studied the environments. Because most people are unfamiliar with leeches, they do not notice the shape and placement of the eyes (which may be enough to indicate to an expert that they belong to different species). However, in the case of fish and the higher vertebrates, which are more familiar to all of us, it is natural to notice different species and give them different popular names. It may therefore seem that there is greater variety among fish populations, but this is not necessarily true. The different kinds of fish—detritus-eaters and alga-eaters, for instance—can be found in almost any eutrophic lake, and those in different countries are often related. It is clearly impossible to refer to them all; but the pattern, as with the invertebrates already mentioned, is usually the same, and we can select examples to illustrate the range of habits and body forms that have evolved.

Although most fish have a skeleton of bone, a few utilize cartilage—the substance around joints in mammalian skeletons—as the major skeletal component. In Lake Nicaragua lives a landlocked freshwater shark that is almost indistinguishable from its marine relatives. Both this shark and the stingrays that can be found in lakes and rivers of northern South America and Southeast Asia have cartilaginous skeletons. The sturgeons and their relatives the paddlefish, of the Northern Hemisphere, have some cartilage, as well as bone in their skeletons. They are both famed as the source of caviar, which consists of their eggs (the roe) with the membranes removed. A single female may lay many millions of eggs; the hausen, a type of sturgeon from the Caspian Sea, yields up to 200 pounds at one time.

Two groups of bony fish are also unusual and may represent surviving members of formerly important groups from prehistoric times that are now largely extinct. The polypterids or bichirs, which can be found only in some African lakes and rivers, are one such group. They have thick diamond-shaped scales and the fins have a stubby shaft reminiscent of that seen among the coelacanths. They can breathe air directly with a peculiar type of swimbladder if the water becomes stagnant but primarily they breathe through gills. The bichir grows to about four feet long and feeds on a diet of small fish, frogs, crustaceans, and other prey.

The second "living fossil" group includes the garpikes and bowfins of North America and they also have a bladder that can be used as a lung. They are called *ganoids* because their thick scales are covered with a layer of ganoin, which is rather like enamel. All the ganoids are predatory fish with strong jaws; these are much longer and thinner in the garpikes than in the bowfins. All of them tend to be rather sluggish and to wait for their prey rather than seek it out. When they capture the prey, it can be firmly held by their batteries of teeth.

The remaining freshwater bony fish can be characterized in many different ways—according to relationships, habits, distribution, and so on. For our purpose let us consider three groups: the bottom-dwellers, the fish that live higher in the water and feed largely on plants or invertebrates, and those that are primarily piscivorous (fish-eating). These groups are by no means sharply defined but will serve as appropriate divisions.

A number of bottom-living fish feed on the invertebrates that inhabit the detritus of a lake bed. Many have feelers or *barbels* (thin tactile organs on their lips), which help them to locate their food because the water is often clouded by suspended particles. The numerous kinds of catfish are obvious examples.

Most catfish live in the lakes of South America, but others occur in Asia, Europe, Africa, and North America. The American catfish (brown bullhead), which has now been introduced to some parts of Europe, is probably the most familiar of these bottom-dwellers; it can be found in North American lakes wherever the water is clear. It and most other types have a flattened underside that enables them to rest on the bottom. And the catfish generally have dispensed with small fragile scales, which could easily be damaged in their habitat. Instead, they have either naked skin with no scales, or a few large bony plates protecting the body.

Two types of catfish can be singled out from the others as differing in some respects. The electric

Armored catfish

These fish are all equipped for life on the beds of lakes or rivers. Mouth probes called barbels *enable the armored catfish of South America to find food hidden in mud. The brown bullhead of North America is a catfish with a tough scaleless protective skin. The clown loaches of Indonesia are less highly specialized. Most curious is the elephant trunk fish from tropical Africa.*

Elephant trunk fish

catfish of Africa is one. Like other "electric" fish, it can produce pulses of electricity—up to 200 volts, according to some authorities—from certain modified muscles in its back. The electricity is used to stun the animals on which the fish feeds, but it is not known to attack men. The electric catfish rarely exceeds two feet in length—which is in obvious contrast to our other example, the European catfish, or wels, which is the largest freshwater fish in Europe. Distributed throughout Asia as well as Central Europe, the so-called European catfish may reach lengths of up to 12 feet and weigh more than 600 pounds. Worms and insect larvae could not satisfy a creature such as this; it feeds on large fish as well as water birds or even smaller mammals.

Other inhabitants of lake bottoms include the loaches and mormyrids. The stone loach of Eurasia is predominantly a stream fish, but there are brightly colored species in Southeast Asia and its offshore islands that are more tolerant of still water. A number of these spend the day buried in the mud of the lake bed, emerging at night to feed on plants or small animals. The mormyrids are the African equivalent of loaches in their habits, but are not in fact closely related. They have thick, slimy skins and an electric organ that works a bit like radar, to help the fish find its prey (the organ is not powerful enough to stun a living creature). Many mormyrids have tiny mouths at the end of strange-looking "trunks," which can be stuck into the mud as the fish grubs about for food.

The second category of fish—those living higher up in the water and feeding predominantly on plants or invertebrates—is probably best represented by the cyprinoids, an enormous assemblage including carp, characins, tench, bream, and numerous others. Many, but by no means all, of this group scavenge for food in and above the mud or among plants, hunting for worms, insects, and snails. The common and the so-called crucian carp are typical in this respect and in the fact that they will eat some plant

Clown (or Borneo) loaches

Brown bullhead

material as well. The crucian carp is closely related to the familiar goldfish, which became popular as an ornamental fish in the Victorian era, and remains so today. The shiner, the mud-eating stoneroller, and the fallfish are related species, which look alike and have similar habits. They are rarely fast swimmers, and do not need to be, because their prey are not fast-moving either. They therefore lack the sleek lines of the fast predators, and often have wide, thick bodies. Their mouths, which tend to be relatively small, lack the sharp, needlelike teeth of fish-eaters.

Nowhere are the above features better seen than among the characins of South America, whose brilliant colors have made so many of them such exciting aquarium fish. Most move about in shoals, feeding on planktonic creatures, but a few, such as the piranha, are able to eat very much larger items. The piranha are of course famous for their voracity. They will attack large animals that enter the waters where they live, and they do so in such numbers that the victim is overcome by the sheer weight. Their razor-sharp teeth enable them to tear their prey to pieces. Other cyprinoids that deviate from the invertebrate diet are the giants of the group. The African tigerfish, and the mahseer and catla of India, which may all reach nearly six feet in length, are just three of a number of cyprinoids that feed primarily on other fish.

One family of fish that are strictly carnivorous but feed almost entirely on invertebrates is the sticklebacks, which are so named because of the free fin rays in front of their dorsal or back fin. They are common—and the quarry of juvenile fishermen—in many parts of the Northern Hemisphere, and most species are freshwater fish. The nine-spined stickleback is most common in America, although it can also be found in Europe; but the three-spined species appears to thrive only in Eurasia. Wherever they live, sticklebacks build beautiful nests in which the drab-hued female lays her eggs after an elaborate courtship by her more colorful mate, who, in the breeding

Above: the thick-bodied, slow-moving bream have relatively small mouths and no jaw teeth. They eat a mixed diet of plants and small animals, including worms and small insect larvae.

The Malayan angel (lower), of the Indian Ocean, thrives in fresh water, and the archerfish (upper) can shoot down insects 20 inches above the surface by squirting water at them.

season, has a beautiful red underside shading to bluish white on top of the fish. When the eggs have been laid the male tends them until they hatch and afterwards until the young can swim powerfully enough to escape his attention.

So much, then, for the fish that are directly dependent on the enormous quantity of invertebrate fauna and that make up the majority of the fish in a lake. From the human standpoint, they are not as spectacular as the third group, which feed on other vertebrates, primarily on the very fish just mentioned. Piscivorous fish are generally large, as they must be to catch their food, and they have batteries of sharp teeth to catch and sometimes tear it. Many are rapid swimmers, using the whole body as well as the fins for propulsion, but others are more stealthy and lie in wait to catch any prey that inadvertently strays too close. A number of the fish-eaters feed on invertebrate fauna when young, and many adults supplement their diet with larger invertebrates.

As already suggested, most of the real giants of the fish world are in the piscivorous group. The European catfish—referred to already as a bottom-dweller—is such a giant piscivore, being the largest freshwater fish in Europe. It is comparable in size and diet to the South American arapaima, which is sometimes said to be the largest freshwater fish in the world. This massive beast, weighing more than 350 pounds and occasionally more than 12 feet long, is a voracious predator. Not only is the adult size impressive, but the arapaima's growth rate is stupendous: as a result of its meaty diet, the young fish reaches a length of up to nine feet in only four years. It may consume anything of suitable size, but perhaps the most common item in the arapaima's diet is its much smaller relative, the arawana. The arapaima pays a penalty for growing so big in that it attracts hungry human beings, for it constitutes a significant item in the diet of many South American Indians, who dry and then eat the flesh. If it escapes the arrows of these Indians, an individual fish may continue to grow; observers have reported specimens 15 feet long.

Another large predator found in the slow reaches of the Amazon and adjacent bodies of water is the electric eel—a creature of the muddy environments where sight is of very limited value. These eels feed at night on fish, frogs, or large invertebrates after stunning them with powerful electric shocks. They use smaller pulses for orientation and searching for prey. The organs that

Above: the pike's jaws easily engulf a tiny stickleback, but stickleback spines prick the pike's mouth and, as can often happen, the pike may be forced to disgorge its prey alive.

The piranha has massive razor-edged teeth. A big shoal of these small fierce Amazonian predators could strip all the flesh from a human body within minutes.

An arapaima in the headwaters of the Amazon. Arapaimas are among the largest of freshwater fish, weighing up to 500 pounds and reaching 15 feet in length. They eat mainly fish.

produce the electricity create a series of weak shocks, which generates an electric field in the eel's immediate vicinity. When prey enter this field the eel can sense them because they disrupt the electrical field. Because of their dependence on these electric means for living in murky situations, the eyes are degenerate in the adult eels. A second adaptation to the environment is their development of air-breathing. Their mouths contain a bed of blood vessels that take oxygen from mouthfuls of air, and their gills are almost useless. If an electric eel is prevented from reaching the water surface, it will drown.

Far more familiar to most of us than the arapaima and electric eel are the perch and pike perch. These carnivores live mainly on small fish, although they may include considerable quantities of invertebrates in their diets. The true pikes and their relatives the mud minnows, which can survive in ice if their body fluids are not themselves frozen, are very well equipped for their predatory existence. They can swim quickly and have elongated jaws as well as a formidable set of teeth. The basses of North America (which have now been introduced to parts of Europe) and many of the cichlids that occur in tropical freshwater habitats have similar features. Other types of cichlid rely on different foods—some feeding on plants, others on plankton, and many on small invertebrates—but it is nevertheless true that a vast number are primarily piscivorous when adult. All these common predatory fish depend on speed and power to overcome their prey, which is often surprisingly large.

Most of the fish noted so far depend entirely on the water to supply their meals, but there are a few species that vary their menu by feeding on items normally outside a fish's range. For example, the archerfish of Indonesia and Australia, which feeds partly on small fish and invertebrates, has the unique ability to shoot down aerial insects. To do so, it squirts a jet of water from its mouth; this hits the insect, causing it to drop onto the water surface. The fish has a surprisingly accurate aim, and large specimens can squirt water for a distance of about three feet. The climbing perch of Southeast Asia has evolved an even more startling method of feeding. Using its gill covers and lower fins as legs, it can leave the water altogether and feed on a wide variety of small terrestrial animals. Like the splendid Siamese fighting fish, the gourami, and the paradise fish, the climbing perch can breathe atmospheric oxygen as well as extracting oxygen from the water with its gills. It does so by using a labyrinth of fine air passages and blood vessels situated above the gills.

A number of other interesting vertebrates form an integral part of many freshwater communities. Many are amphibious, but one unique group is composed of the lungfish, which may well resemble the ancestors of terrestrial vertebrates. As their name implies, the lungfish all breathe air and have one or two lungs for this purpose. The Australian lungfish differs from the others in this group in that it has a single lung and does not bury itself in mud during the dry season. Its African and South American cousins have two lungs and pass the hot dry weather (when the water level may fall considerably) encased in the mud, the African species in especially built mucus-lined cocoons. All the lungfish measure more than three feet in length when fully grown and are carnivores feeding on fish, frogs, and snails and various other invertebrates.

The vertebrates—whether permanent residents of a lake, such as the fish, or visitors, such as birds and mammals—do not necessarily benefit from the invertebrates. On the contrary, a number of the invertebrates actually live at the expense of their vertebrate fellow-residents, this "retaliation" taking one of two forms: the invertebrates may simply eat very small vertebrates in their entirety, or they may feed on a part of a much larger animal without necessarily killing it in the process—a form of existence known, of course, as parasitism.

Anyone who has done much freshwater fishing will be familiar with one example of parasitism—the pathological condition that results from a parasitic protozoan, or *sporozoan. Glugea* is one of many sporozoans that parasitize fish and other freshwater vertebrates. It attacks (or, rather, feeds on) fish, primarily sticklebacks, eventually producing large white swellings on the fish in which there are many tiny spores. These are later released to infect other fish. Because of the resultant rapid multiplication, *Glugea* is extremely widespread in a stickleback population if it exists at all.

It is difficult to generalize about the harm done to its host by a sporozoan parasite, and we can seldom spot their direct effects upon the creatures they attack. This is equally true of some ciliated protozoans that live in the guts of amphibians such as frogs. They do not appear to harm the frog and it is possible, in fact, that they are not parasites at all, but simply inhabitants of a situation where, with a minimum of effort, they get protection and an abundant supply of everything they need. Incidentally, it should not be forgotten that invertebrates as well as vertebrates provide "homes" for protozoans. A number of different types, for example, live on *Hydra*, and others are quite common on the small planktonic crustaceans.

Many fungi are also parasites of fish and their effect can be not only physically obvious but economically harmful to man, for the fungi often infect fish on farms where the fish are being reared for restocking angling grounds. In all cases where the infection is advanced, the fish are severely weakened and die or easily fall victim to predators. Symptoms of fungal disease vary with the actual fungus involved, but you can often see fluffy white masses of fungal filaments on the bodies of diseased fish. Fish that have suffered physical damage, such as the loss of scales or a torn fin, are particularly susceptible to fungal attack, and the fungal filaments are likely to form a conspicuous mass—very frequently at the gill openings, a common site of infection.

Flatworms, which we discussed in an earlier section of this book, are closely related to two important groups of parasites, the flukes (trematodes) and the tapeworms (cestodes). These groups contain representatives that live in or on a great many vertebrates, including man, and many of them are associated with fresh water at some stage in their development. The flukes can be separated into two groups. The first group parasitizes only one type of animal. The second always has two or more hosts in its life history.

Flukes of the first kind inhabit the gills of some freshwater fish, such as minnows. Others occur in reptiles and amphibians. One such single-host fluke takes up temporary residence on the gills of tadpoles, then migrates to the bladder, where it stays after the tadpole turns into a frog.

Among the two or more hosts of the second group of flukes, one element is constant: the first host is always a mollusk, usually a freshwater snail; another host is often a mammal or bird. The sheepliver fluke, as was mentioned earlier, has only two hosts—a snail and a sheep. An example of a fluke that may have three hosts is *Gasterostomum*. This animal starts life in a freshwater mussel, and then moves into a roach. If, as often happens, the roach is eaten by a perch, the parasite is transferred to the perch, and there it stays.

One fluke that is of great and disastrous importance to man is *Schistosoma*. The adult of the species lives in human blood vessels, primarily those of the lower abdomen. The female lays her eggs in vessels around the bladder, and the eggs are equipped with sharp spines that lacerate the vessels, causing a hemorrhage. The flow of blood releases the eggs into the bladder itself. If the unfortunate host then urinates into fresh water, the eggs hatch, and the larvae find a snail for their first host. When mature, the larvae leave the snail and swim freely in the water. Thereafter, if they come into contact with human skin, they rapidly burrow in and enter the blood system, where the new adult develops and the cycle begins anew. Infestation with these flukes results in a debilitating condition called bilharzia—unfortunately a common complaint in many tropical and subtropical areas where poor sanitation prevails.

Tapeworms, or cestodes, have a similar life cycle to that of the multiple-host flukes. Their first host is often a worm or crustacean. The larval tapeworm than transfers to a second, and sometimes a third, vertebrate host. Adult tapeworms always live in the gut, and are frequently found in some freshwater fish. They can be transferred to the guts of birds or mammals. Thus the fish have served as intermediate hosts.

Other worms also live as parasites in vertebrates, sometimes sharing the host's involuntary hospitality with one or more of the kinds we have been discussing. There are many different types of roundworms (or nematodes) to be found in fresh water as well as on land and in the sea. Some of these are free-living, and others are plant parasites; but the remainder are animal parasites. Rather slender, generally tapering at the ends, and creamy white in color, the roundworms have no distinct blood vessels, lungs, or gills, and so they rely on diffusion to meet their oxygen requirements. The parasites among them live at the expense of a great number of freshwater animals, in some cases utilizing two different hosts in their life history, just as do the flukes and tapeworms. Some, for example, live initially in small crustaceans, but move into the tissues of a fish when the crustacean is eaten.

A similar situation involves a nematode known as *Dracunculus*, which lives as a larva in the crustacean called *Cyclops*, or in one of its close relatives. If *Cyclops* is swallowed by a human

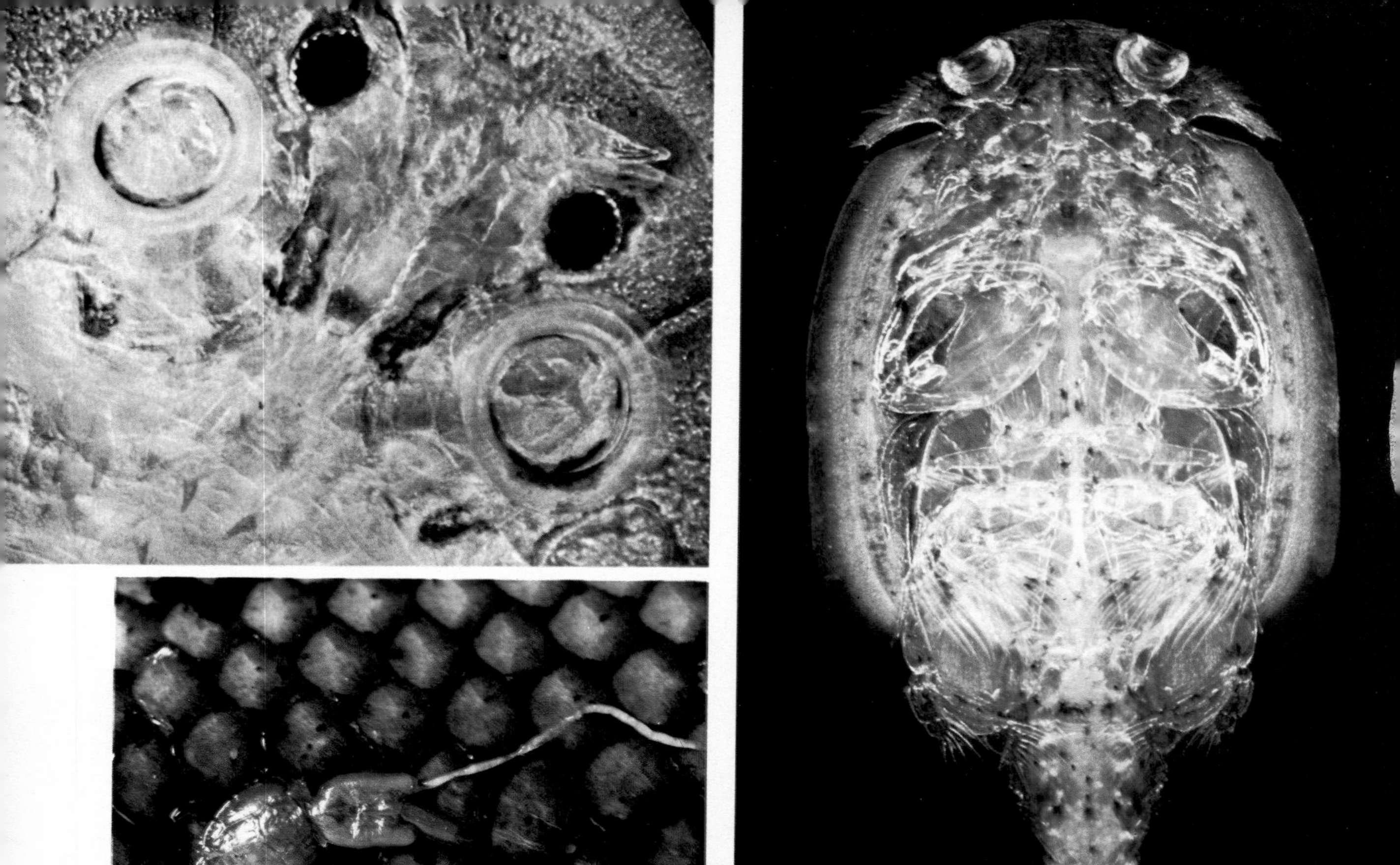

Fish lice are among the most successful of the parasites that prey on fish. Argulus *clings to its host with head suckers (big disks on the magnified photograph, top left). Marine fish lice such as* Lepeophthirus *(above left, enlarged) can be found on the scales of salmon. Above right:* Caligus *fish louse (greatly magnified); its flattened body is designed for clinging to an unwilling host.*

being, the larva is liberated and moves to the surface of the body, just under the skin, where it matures and causes the disease called dracontiasis. The adult nematode may be as much as four feet long.

As we have seen, many leeches feed on invertebrates, but there are also a number that suck the blood of fish and higher vertebrates, including man. The leeches are all *ectoparasites*—that is, they live on the outside of their hosts, unlike most of the worms, which live inside and are called *endoparasites*. The majority of leeches that are parasitic on vertebrates feed on fish, often attacking the soft tissues around the gills. Their activities result in open sores, which may rapidly become infected and lead to death. Ducks and other water birds may also be attacked by leeches, which generally infest the nasal passages or other parts of the body not covered by feathers. Relatively few leeches feed on mammals, but those that do leave nasty open wounds. They secrete anticoagulants, too, which prevent the blood of their victims from clotting. Thus leech sores may continue to bleed for several hours before healing can begin.

Besides the various worms, leeches, and microorganisms, several crustaceans are also parasites of vertebrates. For example, some copepods, which are closely related to those found in plankton, are important fish parasites. Salmon and trout in temperate regions are regularly infected by the copepods. One rather unusual crustacean is the fish louse, *Argulus*, which is probably related to the copepods but has certain distinguishing features. The body is flattened horizontally, and there is a prominent pair of suckers on the underside, with which the louse attaches itself to its victim. When it has found a suitable host, it extracts a meal of blood by means of a proboscis. Usually, the point of attack is the fish's gills, which are soft. The louse's flattened shape minimizes the chance of its being swept away in the water current that passes over the gills. The parasite's relationship with the fish is temporary; the louse can move onto a new host by swimming with its legs and the flexible sides of its body, which undulate like those of a flatfish. Such fish lice are found in both salt and fresh water all over the world.

The wide range of parasites may exercise considerable control over vertebrate populations. When they are considered in conjunction with the invertebrate predators, it becomes increasingly clear that young and small fish in particular must overcome a variety of hazards if they are to survive. It is this battle for survival that has shaped many aspects of fish and their evolution. The fish produce enormous numbers of eggs and fry each time they breed. Many fish lay several hundred eggs every year, and extreme examples such as the common perch lay as many as 200,000. Yet on average only two need to survive to adulthood to maintain a stable population!

We have now seen representative animals and plants of many kinds, all of which—along with many more—make up the lake community. Most of the invertebrates and fish that we have mentioned serve to demonstrate individual life styles in the complex pattern of life in water. And we have nearly come full circle—but only nearly. There remains one more strand in the complex web of life in and around our lake: the living creatures that utilize the last remnants of the energy from the leaf nutrients and the sun shining on algae and higher plants. At each stage, as one animal has fed on another or on a piece of vegetation, it has been unable to utilize everything it has eaten, and has produced urine, which contains nitrogenous waste products and salts that stimulate further plant growth. Even more important, animals produce feces. Raining down through the water is a steady stream of partly digested material, mingling with the dead bodies of plants and animals. The fecal matter that reaches the bottom is mixed with the leaves and twigs, but before this happens some is caught and constitutes at least part of the food of some animals. As examples of these creatures, let us select just three groups.

The first example is the freshwater mussels, which are relatives of the snails and limpets. They are common on the muddy bottom of many lakes, rivers, and canals. There are numerous kinds, ranging in size from several inches down to a fraction of an inch in length. They all have two halves to their shells, which are hinged along one edge, and most of them can move by protruding a muscular foot from the other edge. A sticky trap removes food particles from water drawn into the shell through a tube or siphon. The same water aerates the gills through which the animals obtain their oxygen. If grit enters the shell of some species of mussels, it may be covered with several layers of hard material, thus forming a pearl.

The life histories of the many different species of freshwater mussels show considerable variety. Some produce free-swimming larvae, which may be found in plankton; others produce fully developed young, like their parents but smaller; and some produce larvae that parasitize fish. The large swan mussel has larvae called *glochidia*,

Left: a blood-sucking fish leech feeds upon a young trout that it has killed. The leech clings firmly to the fish's body by means of suckers. Unlike many of the other parasitic worms, the leech feeds only from the outside of its host.

which burrow under the skin of fish and there change into tiny adults while feeding on their hosts. The host fish develops a cyst around the young mussel; the cyst ultimately bursts, and the encapsulated animal is released—perhaps far away from its parents. By this means the mussels are widely dispersed and can colonize new parts of the lake, river, or canal.

The second example of animals that feed on the rain of feces and other suspended matter in the water is the sponge. Sponges are more notable for what they lack than for what they have. They are little more than a crudely organized colony of cells, ramified by channels and lacking an obvious gut or breathing system. In many respects each cell leads its own life, but there is some division of labor. Some cells, for example, produce spicules—rods or spiked spheres of silica—which are characteristic of sponges and form the "skeletons" of freshwater sponges. Bathtub sponges are the skeletons of large marine sponges. The freshwater species are much smaller than their marine relatives (although some of the tropical forms do grow fairly big). In fresh water, sponges are normally found attached to stones or growing on twigs or large plants. Those that live in still water depend on a current of water created by flagella, like those of protozoans, to bring in food, which is ingested by many of the cells in the colony. River sponges can also create currents, but they rely chiefly on the movement of water in their environment. The food itself consists of tiny organisms in addition to the small suspended particles.

Finally we come to the bryozoans, or moss animalcules. These are another sedentary group of animals, but they have a far more elaborate internal structure than the sponges or than *Hydra*, to which they show some superficial resemblance. There are at least two distinct types of animal with a number of characteristics in common in this group. All of these tiny creatures are anchored at one end to leaves, plant stems, exposed roots in shallow water, or stones,

Swan mussels lie partly buried in a pond bed, carefully angled so as to draw in water and food particles and expel wastes through openings in their hinged shells.

and bear at the other a horseshoe-shaped crown of tentacles. In some bryozoans, the so-called *ectoprocts*, the mouth opens into the crown, but the anus is outside the crown near the tentacles. In what are known as the *endoprocts*, both the mouth and anus open among the tentacles. The body of the animal has a definite form, unlike the sponges, with an obvious gut, gonads, and a nervous system with numerous fine nerves.

The great majority of bryozoans are colonial and form an encrusting sheet in places that are suitable for their existence. A few, while still anchored, are able to wave about slowly because they have flexible joints in the trunk. Food, in the form of diatoms, protozoans, and small particles (including fecal matter), is collected by the tentacles and passed into the gut for digestion. Bryozoans are extremely successful animals and are widespread in freshwater environments.

We have now completed a circle and are back at the beginning, with mussels, sponges, and bryozoans feeding, at least in part, on suspended detritus (including waste matter from other animals) and re-forming some of its components into living tissue. When this living material dies, it returns to the lake bottom to serve as food for the microorganisms and other animals with which we began. It should now be apparent that wherever we had begun our look at the fauna and flora of lakes, we could have continued around a cycle and finished where we started.

No lake is, however, totally self-contained. From outside our lake come the leaves, which, as we have seen, are an important factor for life in the water; and also from outside comes the sunlight that enables the plants to make sugar and other components essential for life. Substances washed into our lake from surrounding lands are also important, because such substances can tilt the balance of any lake community in different directions. In an earlier section of this book we looked at the different kinds of lake, and this information allows us to make some predictions about the type of community to expect in a given body of fresh water. Intelligent use of such knowledge, along with a survey of some of the key groups in our cycle, such as the planktonic algae, enables us to deduce a great deal about the probable life in a particular lake.

You might expect a lake to become choked with plants and animals, because energy in the forms that we have noted is continuously entering the water. But this, of course, does not happen. Not only do some organisms escape through rivers or streams entering and leaving the lake, but other organisms are removed by temporary visitors to the water. Moreover, one major use of energy is in simply maintaining life itself. The sheer effort of moving about, of reproducing, and of keeping the body "organized" by such measures as the replacement of worn parts utilizes a vast amount of energy that can be derived only from food. Plants can be said to make their own food, but animals must eat. The dissipation of energy must balance the input, or the system would collapse; and it is this that determines the form of the community in terms both of variety and of the absolute number of organisms that can survive.

Some generalizations can be made. Each predator must eat many prey in its lifetime, and each victim requires several items in its diet if it is to grow and survive until eaten. Therefore there will be fewer large animals than small—a principle that extends throughout the community. For every large predatory fish there are many smaller ones, which in turn eat more herbivores, which eat vast amounts of plant tissues. Some parasites, such as the fish lice, may complicate the issue, because several can feed on one fish; but the principle applies wherever large animals eat smaller ones.

The lake is therefore a very delicately balanced environment, containing many interdependent forms. Over short periods of time the elements that make up the system do not change markedly; as we have seen, there is a continuous cyclical process in operation. But evolution of the lake types does result in a gradual change in the proportion and nature of the elements. These changes must take place slowly as newer populations build up or old ones decline.

Food Web for a North American Meadow Pond

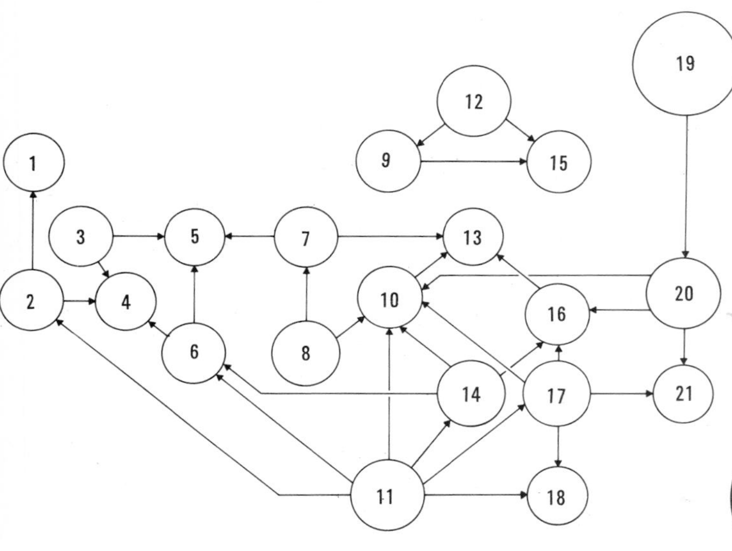

1 Duckweed	12 Bull frog
2 Great pond snail	13 Motile green algae
3 Lesser water boatman	14 Dragonfly nymph
4 Sessile algae	15 Drowning fly
5 Diatoms	16 Mosquito larva
6 Tadpoles	17 Backswimmer
7 Rotifer	18 Water slater
8 Hydra	19 Green heron
9 Water strider	20 Catfish
10 Water flea	21 Tubifex worms
11 Great diving beetle	

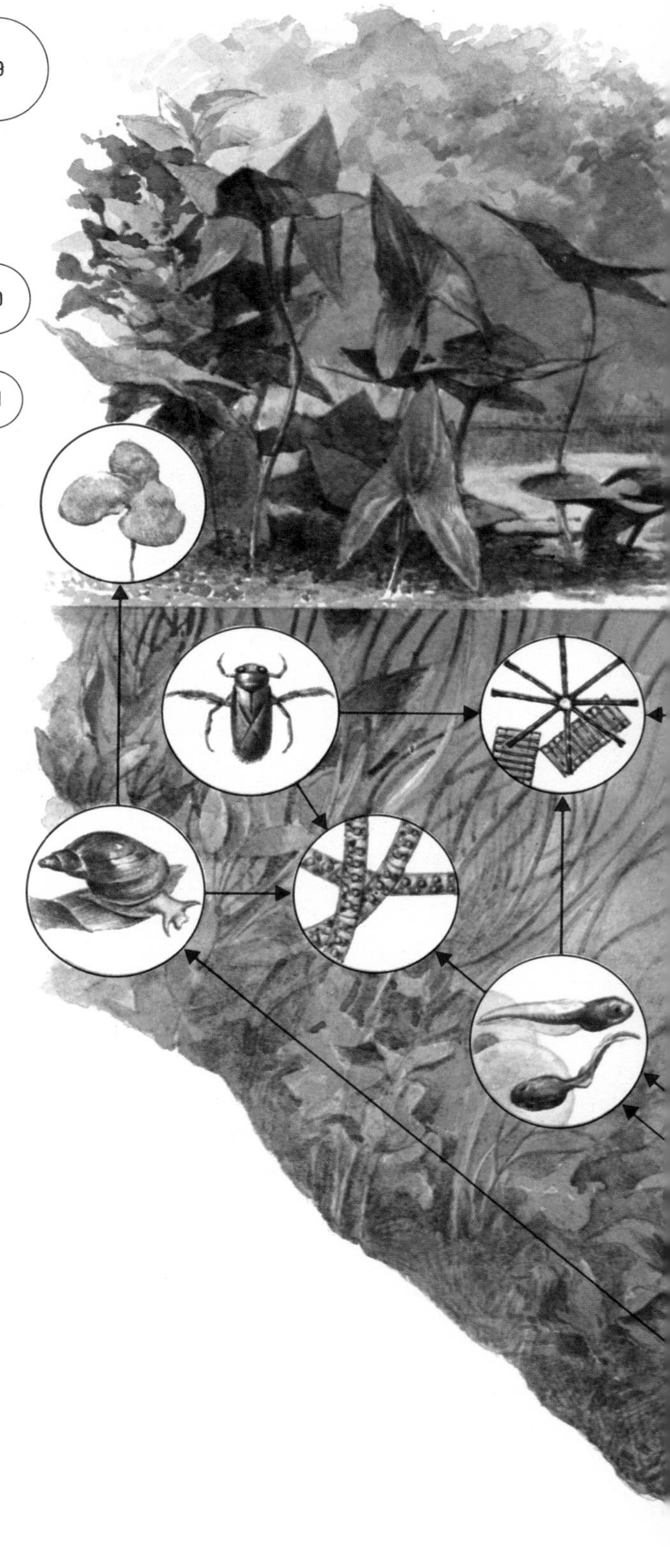

The plants and animals in a pond depend directly or indirectly upon one another as a source of food. We may think of the invisible links thus forged between individual kinds of organism as the threads of a great web. This illustration shows some of the main threads that make up the food web of a meadow pond in North America. Each thread bears an arrowhead pointing from eater to eaten throughout the web. (Plants and animals are not shown to scale.)

Starting with duckweed (1), we notice that this small plant provides food for the great pond snail (2). The pond snail also eats sessile algae (4). But sessile algae may also be consumed by the lesser water boatman (3). The lesser water boatman's diet includes tiny diatoms (5). Diatoms are trapped too by various kinds of rotifer (such as 7). In addition to diatoms, rotifers eat motile green algae (13). The water flea (10) preys upon motile algae and in turn falls victim to larger creatures including dragonfly nymphs (14). The mosquito larva (16) provides food for the dragonfly nymph and the much larger catfish (20). The catfish may itself end in the gullet of a green heron (19). Other arrows connect most of these organisms with at least one other. Even the seemingly isolated surface group of bullfrog, water strider, and fly (12, 9, 15) is really linked with the main pond web, for the bullfrog began life as a tadpole (6).

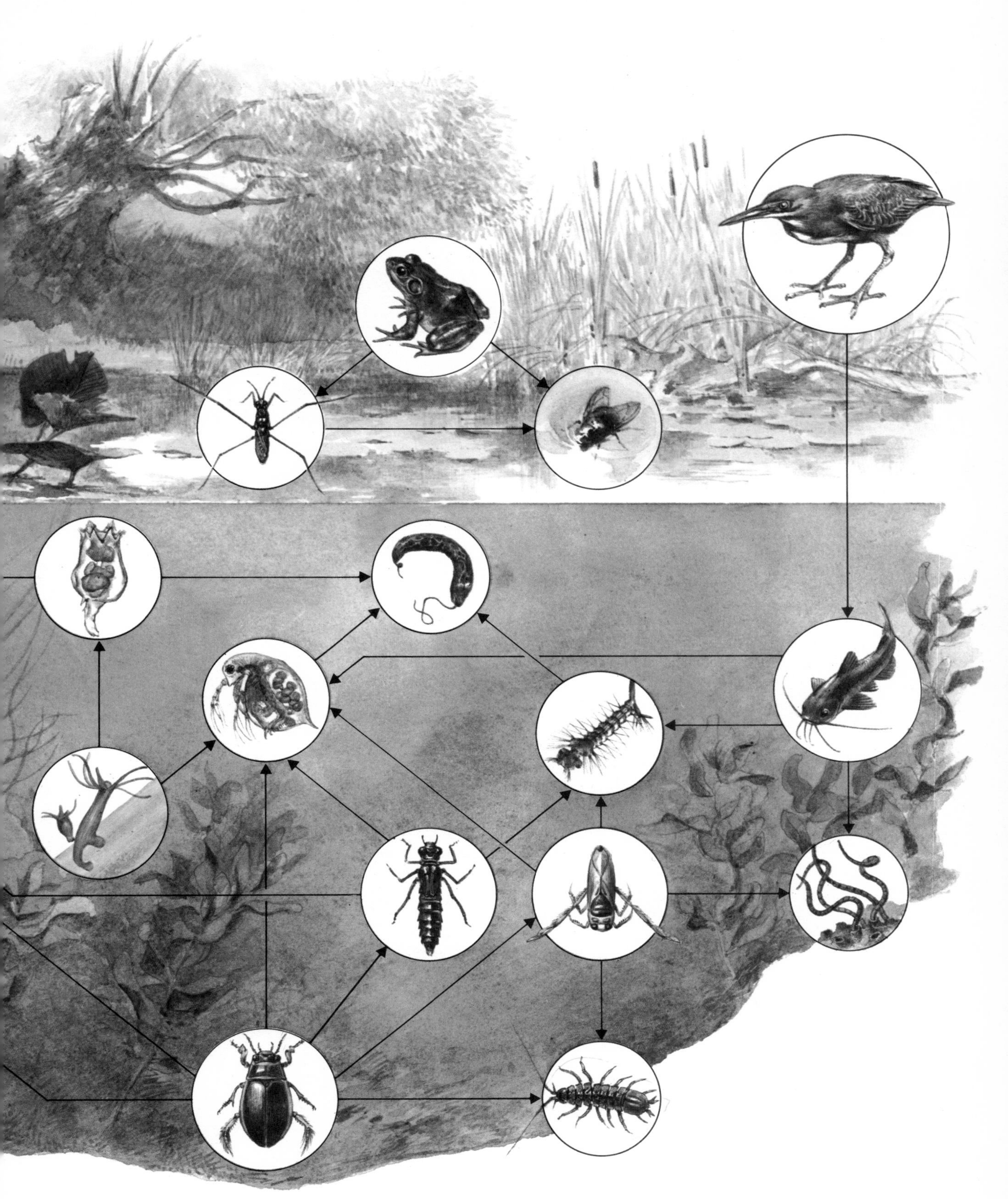

The River Community

Many living organisms found in flowing water are at least partly dependent for their survival upon the bottom substratum, the form of which is determined chiefly by the speed of water flow. Because of this, the speed of flow also largely determines the makeup of both plant and animal communities in each section of a given river. In the stretches of a river where silt is deposited and the flow of water is at its slowest, the communities tend to be very much like those of eutrophic lakes. But in sections with more rapid flow they are distinctly different. This is due not only to the influence of water movement, but also to the water being colder because it cannot absorb heat as standing water can, and to it containing more oxygen because the turbulence exposes more water to the oxygen in the air.

These properties have predictable effects on the flora and fauna. The organisms must either swim powerfully or be securely anchored to prevent being washed away; or else they must manage to escape the force of the current by hiding under stones or by building themselves shelters. The continual expenditure of energy required for the battle against being washed away may be part of the reason why many river species need the large amounts of oxygen carried by flowing water.

Other characteristics of river life are best seen at the community rather than the individual level. Plankton and floating plants are largely absent, for the current removes them from all but the very slowest or stagnant parts of the water course. Because of the current, too, river creatures have rather different reproductive patterns from lake-dwellers. As you may remember, many still-water animals produce small larvae or eggs that drift about in the surface waters. But when water is in continuous motion, as in a river or stream, the eggs must be deposited under gravel or stones, or must be attached to some such solid, immovable object as a rock or large leaf. In some cases, in fact, the eggs hatch out and the young develop inside the river-dwelling mother, and

Sleek, powerfully muscled bodies and broad, oarlike tails thrust salmon up rapids to their spawning grounds. Many river animals must be strong swimmers or tenacious crawlers.

never pass through the free-living larval stage.

Fortunately, there are benefits as well as restrictions in water movement. For example, if some means of trapping it can be devised, food is being continuously provided by the current. Thus, for filter-feeding creatures life may be somewhat easier than in a lake. The flow may also assist dispersal. Large numbers of some species "commit" themselves regularly to the current and drift downstream. These movements are deliberate, not accidental; they occur at certain times of day or during specific seasons of the year, and the effect is obviously dispersal of the species.

There is one further fundamental difference between river and lake communities. The absence of elements of a lake, such as plankton, restrict the variety of feeding habits available in a river. Consequently river communities are frequently less complex than those in eutrophic lakes. Furthermore, because flowing water distributes nutrients and organisms downstream, it is not essential for every occupant of a food web to be present over a short stretch of the river. Predators in one section may be able to feed on herbivores that have been washed into their habitat from higher in the water course. For this reason we should not study isolated sections of a river without due attention to adjoining ones.

We can therefore conclude that there are two essential differences between the flora and fauna of lakes and those of streams or rivers; at any one point in a water course—barring a slow-moving

Above: three "tail" filaments and projecting abdominal "gills" help to reveal this (enlarged) insect as a mayfly larva.

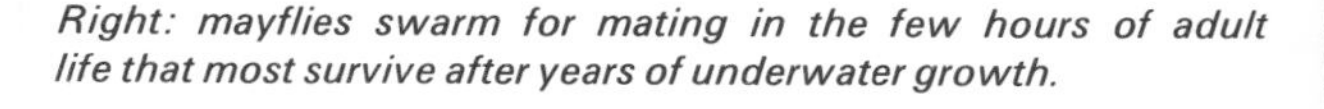

Right: mayflies swarm for mating in the few hours of adult life that most survive after years of underwater growth.

section where silting is heavy—there are likely to be fewer species than in a lake; but over the entire river system there is an immense variety of organisms to match the differing kinds of available habitat. In the light of this understanding, let us look at some of the animals we can expect to find in a river or stream, remembering as we do so that the rate of flow and the character of the bottom will have a strong influence on which species actually occur in a given section.

Many of the groups of invertebrates that we saw in our typical lake have representatives living in flowing water. Beetles, water bugs, *Hydra*, flatworms, leeches, and many others are all found in suitable localities and generally they play similar roles to those that we observed in the lake. Let us therefore concentrate on those additional animals that illustrate specific points about life in rivers.

Of these the larval stoneflies and mayflies are among the most outstanding. Both larvae and adults of almost all stoneflies can be recognized by two long filaments extending backward from the tip of their abdomen (a very few adults lack this characteristic feature). All the larvae have two claws at the end of each of their six limbs; it is with these that they cling to stones in the faster currents. The larvae cannot swim strongly, and spend most of their time creeping about, pressing their bodies against the surface over which they are moving to prevent themselves from being washed away. Of those larvae that

Above: food in the form of a dead fish tempts a caddisfly larva partly from its protective mobile home, made from bits of leaf. Adult caddisflies (below, mating) are drab brown insects. Caddisfly and mayfly larvae both occur in rivers.

must pass the winter in the water many burrow into the stream bed for safety and survival.

As the stonefly larva approaches maturity, its adult wings begin to develop as small buds. Ultimately, the larva is ready to molt for the last time and become an adult. It then climbs on to a rock, casts its larval skin, unfurls the wings, and flies away. Having spent a year or so as a larva, it has only a brief adult life, lasting from a few hours to a day or so. After mating, the females return to the water to lay their eggs, either dropping them while flying over the water or occasionally dipping their abdomens into the water, thereby depositing the eggs. The stonefly larvae breathe through hairlike gills, which may appear as hairy tufts on the larvae of the larger species or be few and fingerlike on those of the smaller ones. These gills are found on the thorax and at the tip of the abdomen. The larval diet varies according to species. Some are carnivorous, feeding on worms and insect larvae, others are vegetarian and browse on the algal coating of stones, and the rest consume the organic detritus that accumulates in pockets behind or beneath eddies.

Mayfly larvae bear a rough resemblance to some stonefly larvae, but they invariably have three filaments attached to their abdomens, from the sides of which also arise tufts of hairlike gills. There is a unique stage in the life history of the mayfly, called a *subimago*—this is a dull, winged adult that emerges at the water surface from the mature larva. The subimago soon settles on shore near the water, casts its "skin," reveals the familiar, shining adult that most of us recognize as a mayfly, and then flies away to mate. Like the stonefly it returns to the water to lay its eggs. The name *mayfly* is derived from the fact that many species leave the water in great numbers during the late spring and early summer.

The mayfly larvae display a variety of life styles. A few species live in lakes or ponds, but most of them prefer running water. Some burrow into the mud or sandy bed of streams and small rivers or live in the banks. There they build tubes in which to live, and most of them feed on detritus and microorganisms around them. Other species also live in the mud or sand but do not build tubes; they creep about on the bottom or on water plants and are primarily carnivorous. Very different are the species that cling to stones in the faster parts of flowing water. They are markedly flatter in shape than the others, and even their legs are compressed, so that the animal can get a firm grip on its stone without protruding into the current. The three tails, or *cerci*, are usually very long in these larvae. Finally, there are the mayfly larvae that swim about freely in the open water. As you might expect, they are found in slower-flowing water than the others, and their slim bodies have long, hairy cerci with which they propel themselves along. Their rather slender legs are not used for propulsion.

Among our list of the most noteworthy inhabitants of running water we can include the caddisfly larvae. They superficially resemble some beetle larvae, with their stout head capsules, small antennae, and biting mouthparts. Unlike the beetle larva, though, only one or two of the caddisfly larva's three thoracic segments is partly toughened, and the remaining segments look like those of the abdomen. Caddisfly larvae have three pairs of legs, of which the first—those nearest the head—is always the shortest. A distinguishing feature is a pair of hooks on the tip of the abdomen; these can be used for gripping the surfaces over which the animal moves or for securing it inside the case that many of these creatures build for themselves.

The many different kinds of caddisfly larvae can be divided into two distinct groups on the basis of their body form and way of life. Some, which are large, rather bulbous creatures, generally live in tubelike cases built from pieces of leaf, sand, or other elements in the detritus. Most of these case-dwelling larvae seem to be vegetarians—but we are not entirely certain about the diet of many of them. This group can be commonly found in lakes and ponds as well as in slow-flowing water.

The second group have slimmer bodies and are very much more active. Some of them build themselves cases made entirely of silk, which—like the silk secreted by many other insects—is actually a special kind of "saliva" produced from glands in the animal's head. Many of the species in this group, however, do not have cases, but spin nets between plants or rocks in which to catch their prey. These nets, which are sometimes quite complex, are usually laid across the prevailing current to trap animals drifting downstream. Obviously, then, the net-builders are found only in running water, not in lakes or ponds. These active larvae of the caddisfly appear to be chiefly carnivorous.

The life cycle of caddisflies is sometimes quite long. In temperate regions it often lasts a year or

Above: pincers raised, a female crayfish protects the babies clinging to her belly. Crayfish frequent clear, shallow streams.

Above: still-water snails must surface periodically to breathe, but this snail, found in running water, extracts oxygen from water by means of its gill.

more. Whether it has lived in one or not, the larval caddisfly always pupates in a case. It may be modified from the larval shelter or built especially for the purpose, and it commonly provides a home for the winter. The pupa leaves its shelter when the adult is ready to emerge. It rises to the water surface, sheds the pupal skin, and flies off. The adult itself is usually a dull brown or gray color and has long antennae and thin wings, densely covered with hair. You can easily confuse some species with moths if you catch only a brief glimpse of them.

The larvae of numerous true flies are also to be found in running water. Many of them are closely related to the midges or mosquitoes that inhabit freshwater lakes. One of these groups—of particular (and unpleasant) importance to man, because some species are carriers of disease—is the blackfly group. Their larvae attach themselves to stones, in the fastest portion of the stream, by means of a special pad at the rear of the body. They then collect suspended material and algae with a pair of brushes that whisk particles into the mouth. One unusual adaptation to the environment is the larval blackfly's ability to produce a "lifeline" and attach it to its stone. Should the larva then be washed off, it can climb along the line to regain its former position.

Larval blackflies are not in themselves a menace to human beings, but the adult female is an active bloodsucker and may bite people in order to feed. In so doing, particularly in the tropics, she may transmit diseases such as onchocerciasis, which can result in blindness. Even when the consequences of a bite are not so disastrous, the bites themselves may produce severe inflammation. This type of inflammation is often diagnosed incorrectly as having come from a midge bite. The swarms of tiny flies near running water are frequently blackflies, not midges. There are, to be sure, a number of midges that live in both still and running water and that do bite people, but they are so very tiny that they are sometimes not even recognized as flies.

Other than the insects, certain crustaceans are also characteristic of clean, flowing water, including freshwater shrimps of the kind already described. In a few places there may be some true shrimps—true in the sense that although much smaller they are closely related to the edible marine species. There are also a number of freshwater crabs, looking very like the seashore varieties, and crayfish, which resemble small lobsters. Crayfish are the largest freshwater crustaceans. Although they are relatively scarce in many countries, North America can boast a great many species—more, in fact, than in all the rest of the world. Crayfish have a pair of large *chelae* (pincers) with which to catch the small animals that make up the bulk of their food; some species may take a little plant material as well. The many different families of crayfish show great variety in size. Some are only one to two inches long, but others are six inches or more. The largest species take perhaps four years to reach maturity, and some have been known to live for up to 20 years.

Snails, too, are abundant in flowing water. Unlike the still-water snails, they are predominantly operculates—that is, they can close the entrance to the shell, and they have gills rather than the lung that characterizes lake-dwelling pulmonates. They lead a similar life to that of the pulmonates but do not need to rise to the surface to get their oxygen. Their inability to use atmospheric oxygen may be something of a handicap to their distribution, but their ability to seal the shell does permit them to survive adverse conditions, at least for short periods.

The freshwater winkles are probably the most familiar of the operculate snails. They are large snails with shells that may be over one and a half inches in height, and are brownish yellow in color with three dark bands running around each whorl. As in most other operculates there are separate males and females, but, unusually, the females produce fully developed young instead of laying eggs.

The various insects, crustaceans, and operculate snails are just a few examples of the invertebrate groups associated with running water. But it must be remembered that the descriptions we have made of these animals are generaliza-

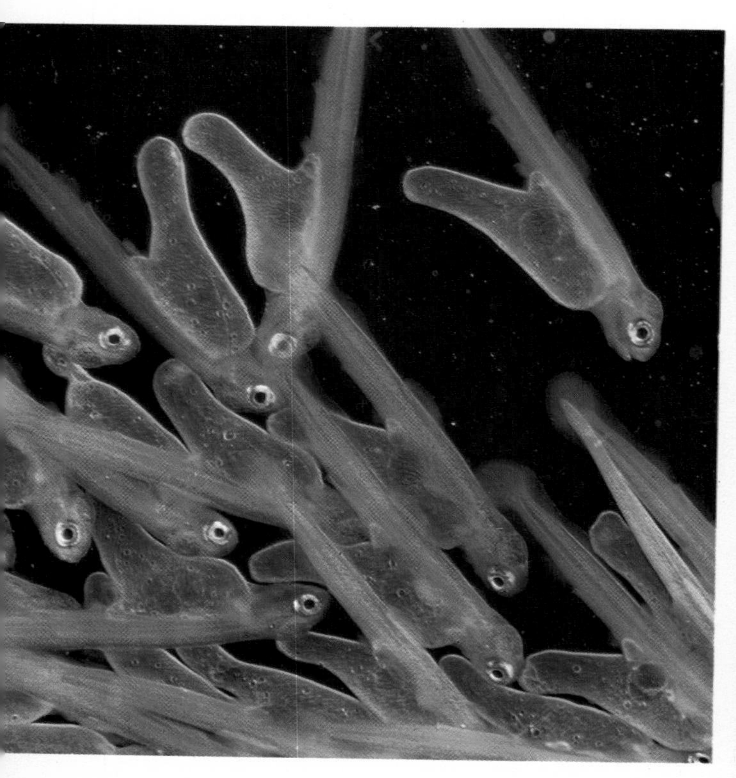

Above: yolk sacs projecting like yachts' keels provide these salmon fry with all the nourishment they need. When they have absorbed their yolks young salmon squirm up through the gravel beds in which they hatched, and hunt for food.

Right: Kokanee salmon congregate to spawn in a shallow mountain stream feeding Lake Tahoe. Unlike other Pacific salmon, which mature at sea, this race of sockeye salmon reaches adulthood in lakes, and spawns in their tributaries.

tions that cannot possibly cover the multitude of special cases. The groups that we have examined are by no means exclusively limited to the particular environments in which we have looked at them; we have had to oversimplify in order to avoid endless complexities. This holds equally true for fish; one species of fish may be found in a variety of habitats. We shall discuss here only those groups that are found more commonly in rivers than elsewhere.

Salmon, as most people know, are extraordinary in being river *and* sea fish. They are spawned in the headwaters of rivers, where they remain until they are young adults, after which they move downstream and out to sea to join their elders. In time they return to the rivers to spawn, thus perpetuating this cycle. The sea lamprey also enters rivers to spawn, otherwise living largely in the sea, but there is an important difference between the two groups in the selection of spawning grounds. The salmon spawns in clean water over a gravel bed and the young feed on egg yolk in their yolk sac until it is all used up; by this time they are big enough to hunt for other food. The lamprey, on the other hand, spawns on a muddy bottom, because its young—known as *ammocoetes*—begin their feeding on the detritus found there. Like the salmon, the ammocoetes remain in the river until they have matured into the adult condition, when they enter the sea to begin feeding parasitically on fish. There are a few species of both salmon and lampreys that spend their entire lives in fresh water.

Eels spawn in the sea but spend most of their life in fresh water. Other adult fish frequently encountered in fast streams or rivers include the trout, only a few species of which ever go to sea, the grayling, and some of the pike perch, such as the walleye. As a general rule, all of these fish are dependent on cold, oxygen-rich water for

survival; many die if subjected to a warm-water environment. And they are almost exclusively carnivorous, feeding on small animals in the water and on those that become trapped on the surface, such as flying insects.

One group of fish that overcome the problem of breeding in fast-flowing water in an unusual manner are the so-called livebearers. Unlike the many species that bury their eggs or stick them onto suitable surfaces, the livebearers (as the name suggests) give birth to fully developed baby fish. At birth, the young can swim quite strongly, have lost the yolk sac on which most newly hatched fish depend, and actively search for their food. Included in this group are many of the most popular aquarium fish. The guppies, renowned for their magnificent colors and flowing tails, are descendants of small, inconspicuous stream-dwelling fish from Central America. The swordtails, platyfish, and mollies can also be seen as rather dull green or brown creatures in Central American streams; from them have sprung many of the brilliant red, yellow, or black animals that grace so many fishtanks today.

From the examples we have just considered, we can see that some river creatures are a strange mixture of animals that have retained close associations with the sea either in their behavior or in their structure, and those that have evolved from terrestrial stock but are now beautifully adapted for the totally different life in running water. The salmon and the eel are among the most obvious cases where the association with the sea has been maintained, but animals such as the operculate snails show similarly striking links with seashore animals such as the periwinkles. The stonefly, mayfly, and caddisfly larvae are typical examples of terrestrial origin.

One other group is, surprisingly, mammalian: the river dolphins. These animals are found in a number of tropical rivers, among them the Amazon, Yangtze, and Ganges. They are permanent residents and should not be confused with the whales or dolphins that are occasionally sighted in other freshwater localities. They are smaller than their marine cousins, reaching sizes of five to eight feet, and they can spend only a short time—up to 30 seconds—under water without breathing. It has been suggested, however, that the early evolution of the cetaceans actually took place in fresh water, and only later did they enter the sea. This would account for the true freshwater dolphins still retaining a number of characteristics that may be regarded as "primitive," and they certainly lack some of the fine specializations shown by their marine relatives.

Thus the river, like the lake, is a habitat full of variety. No two rivers are identical. The animals and plants are governed by an endless struggle to survive. They show changes and differences that are related to their total structure. Geographical location, climate, topography, and geology are interwoven in complex patterns that determine the kinds of community that survive. Taking such features into account, it is not surprising that freshwater life is neither totally predictable nor completely incomprehensible. Patterns of adaptation can be recognized, and so can the interrelationships of organisms, whether for mutual benefit or in the benefiting of one from another. Together, all these complex patterns and interrelationships have produced the wonderful living worlds that we have examined.

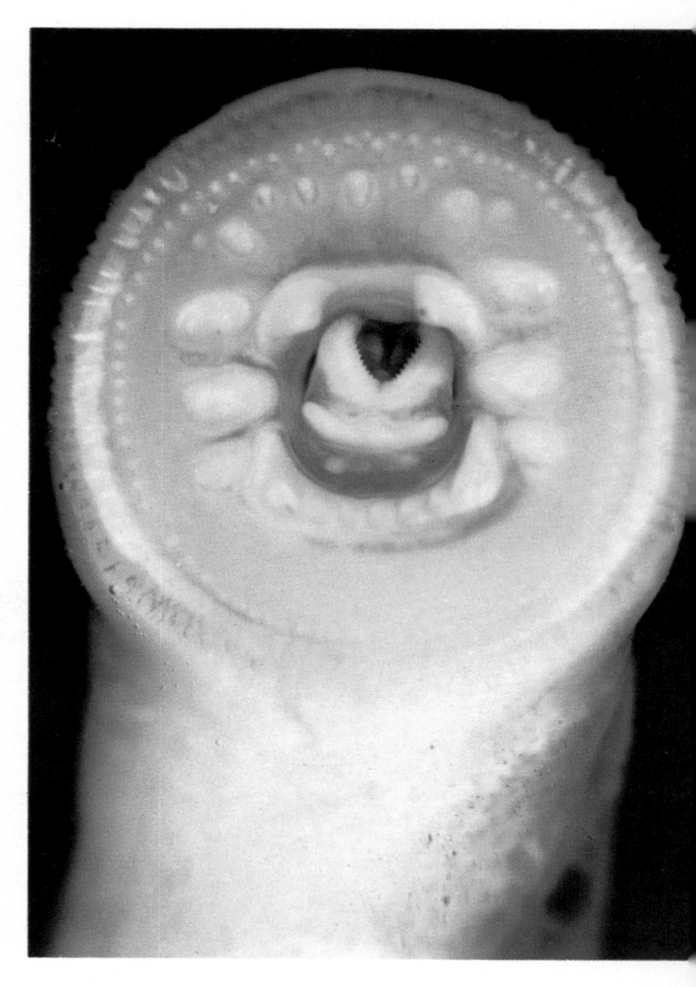

Right: transparent baby eels, called elvers. *This is how the eels look on reaching European rivers after a three-year journey from the Sargasso Sea near Bermuda, where they hatched from eggs laid some 20,000 feet below the surface. The eels mature in European ponds and rivers, then swim back to the Sargasso Sea, spawn and die.*

Below: a lamprey firmly attached to a trout by its mouth. Below left: a close-up view of the lamprey's disk-shaped mouth shows its horny, rasping teeth. Like salmon, lampreys usually start life in rivers, migrate to the sea, then return for breeding. But unlike salmon, adult lampreys are entirely parasitic.

Food Web for a Typical British Upland River

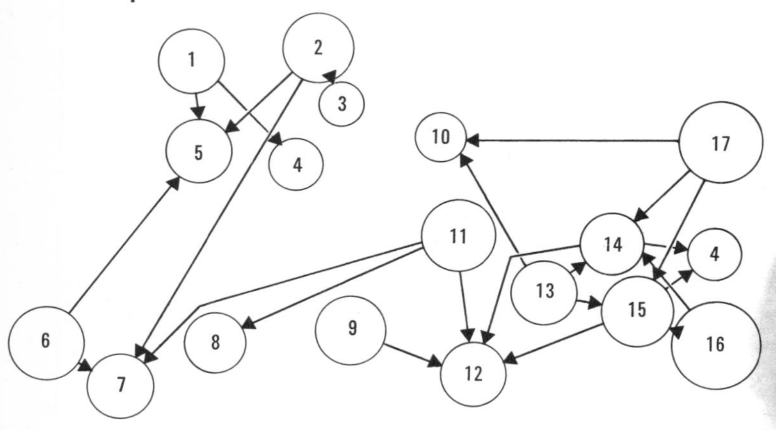

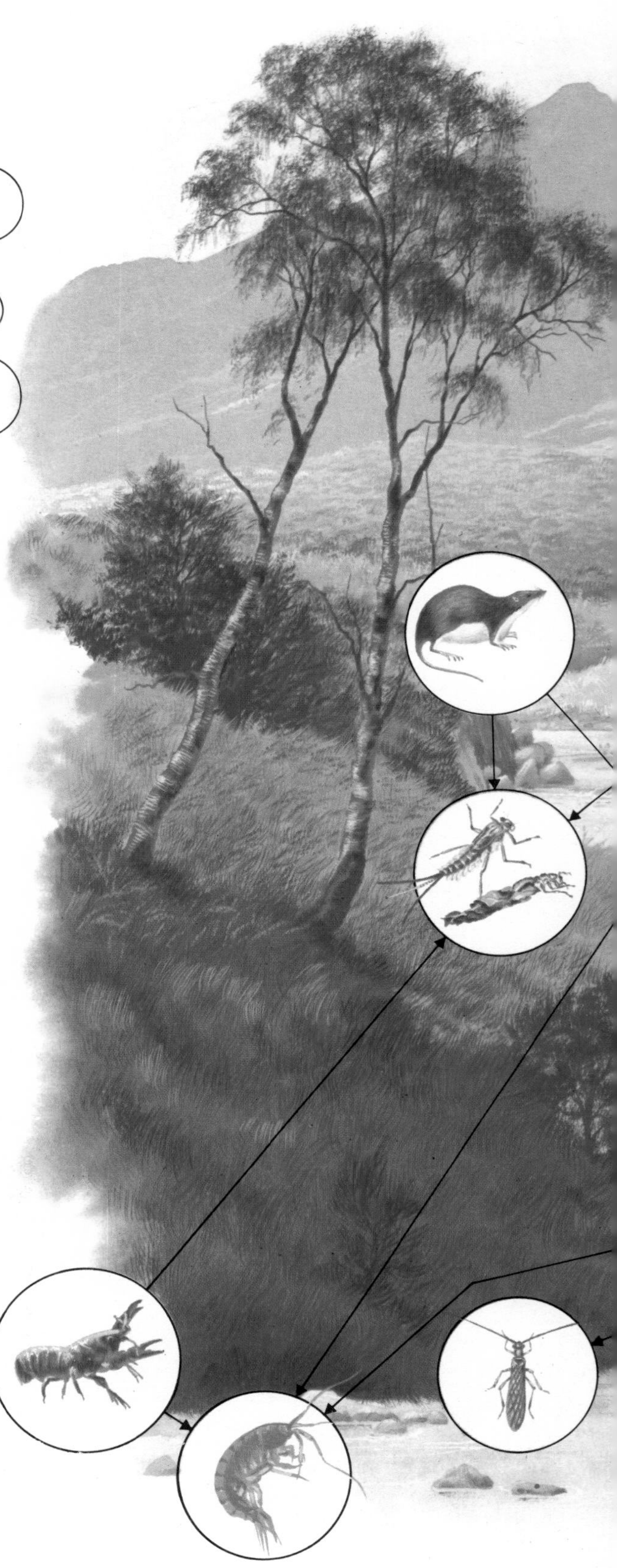

1 Shrew
2 Dipper
3 Stonefly larva
4 Limpet
5 Mayfly and caddisfly larvae
6 Crayfish
7 Shrimp
8 Stonefly
9 Wagtail
10 Eel
11 Sandpiper
12 Mayfly and caddisfly
13 Merganser
14 Trout
15 Salmon
16 Otter
17 Heron

Many rivers in the west and north of Britain flow swiftly through steep mountain valleys to the sea. The force of water prevents mud settling on the riverbed. Thus large plants find no roothold, and no floating or fragile plants become established. Lack of such plants and the sheer force of running water largely give the food web of an upland stream its special character.

Among the stream's small invertebrates (shown here not to scale) determined crawlers find food or cover among stones, or mosses clinging to the stones. Such creatures include the stone-fly larva (3), mayfly and caddisfly larvae (5), freshwater limpet (4), and shrimp (7). These animals provide food for ones that can hunt in rapid currents—creatures such as the water shrew (1), dipper (2), crayfish (6), trout (14), and salmon 15).

Insect larvae that become winged adults leave the water and some furnish food for animals that feed largely at the water's edge or at its surface. Thus adult stoneflies (8), mayflies, and caddisflies (12) fall prey to sandpipers (11), gray wagtails (9), trout, and salmon. In turn, the river fish are sought by predatory birds and mammals, especially the red-breasted merganser (13), heron (17), and otter (16).

The food links we have just described bind all the pictured animals together. But our illustration actually simplifies the food web. Other strands might well be added to it. For example, the water shrew—a tiny but fierce predator—attacks crayfish and fish as well as small invertebrates.

On and Around the Water

All animals need water, and most must obtain it by drinking. As almost every terrestrial animal can periodically be found near fresh water, we shall concentrate only on those animals that spend more time near water than is required merely by occasional thirst. There is no real dividing line between the occasional visitor to fresh water and those that are more closely associated with it, but we can find patterns of activity that have special correlations with water.

Let us begin with the creatures that occupy an intermediate position between those that live in water and those that live around it. We are all aware that the surface of a body of water can act as a physical support—to the emerging mosquito, for example—or as a trap, if it is partly penetrated. Indeed, we are so familiar with some of the properties of surface tension that we seldom notice them. We quite often see things such as hairs, dust, leaves, or paper floating on water, and then sinking as the object becomes saturated with water or the surface tension is broken by the addition of such agents as soap. It is only the phenomenon of "wetting" that traps insoluble material in water.

Many kinds of animals rely on the surface tension for support, and spend their entire lives in this situation, feeding on trapped prey and detritus. An animal that lives on the water surface must be small; and, unless it is really tiny, its weight must be spread over a considerable area to keep it from falling below the surface. The delicacy of the situation is such that if these creatures become wet, they will themselves be as hopelessly trapped as the prey they feed on.

Considering the requirements of this group, it is not surprising that most of the larger surface-dwellers are insects. The most familiar of them are probably the pond skaters (also called water striders). The many kinds all have fairly slender bodies and relatively long legs that spread the insect's weight. They are predators and feed on

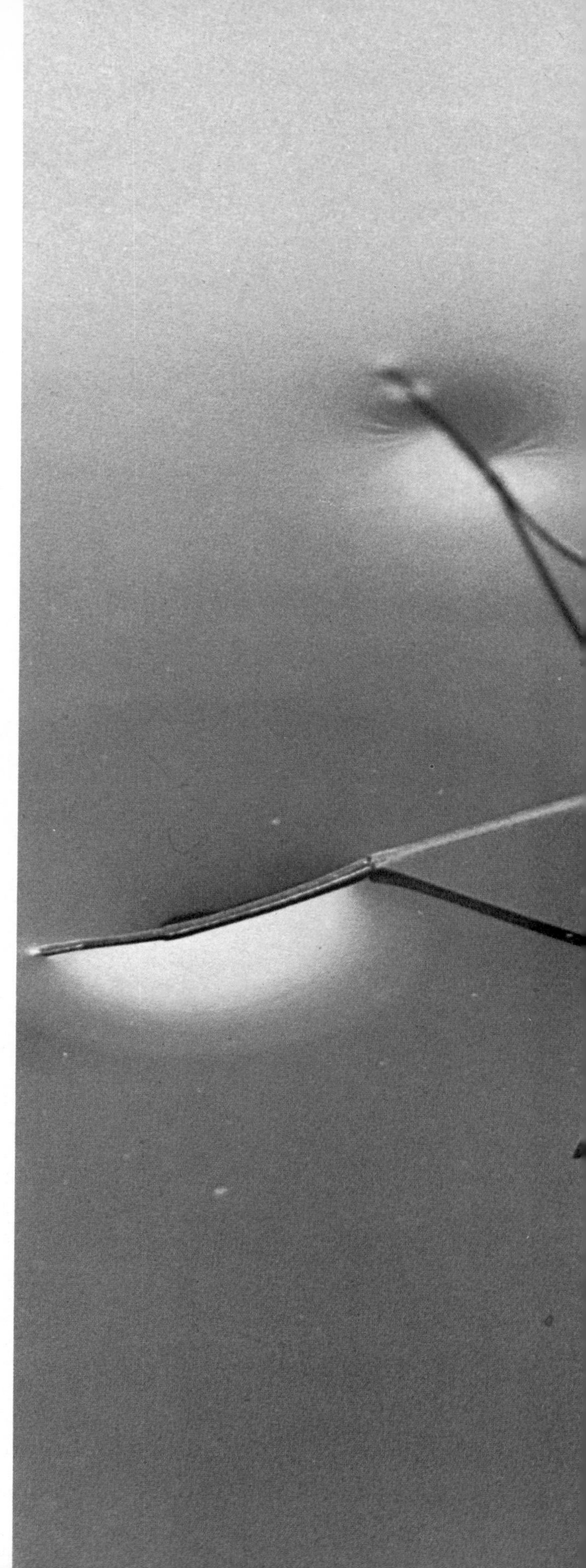

The half-inch pond skater leads a precarious life upon the water surface. Only straddled legs that spread its weight and an air-cushion of tiny hairs stop this bug breaking through the surface "skin" and drowning.

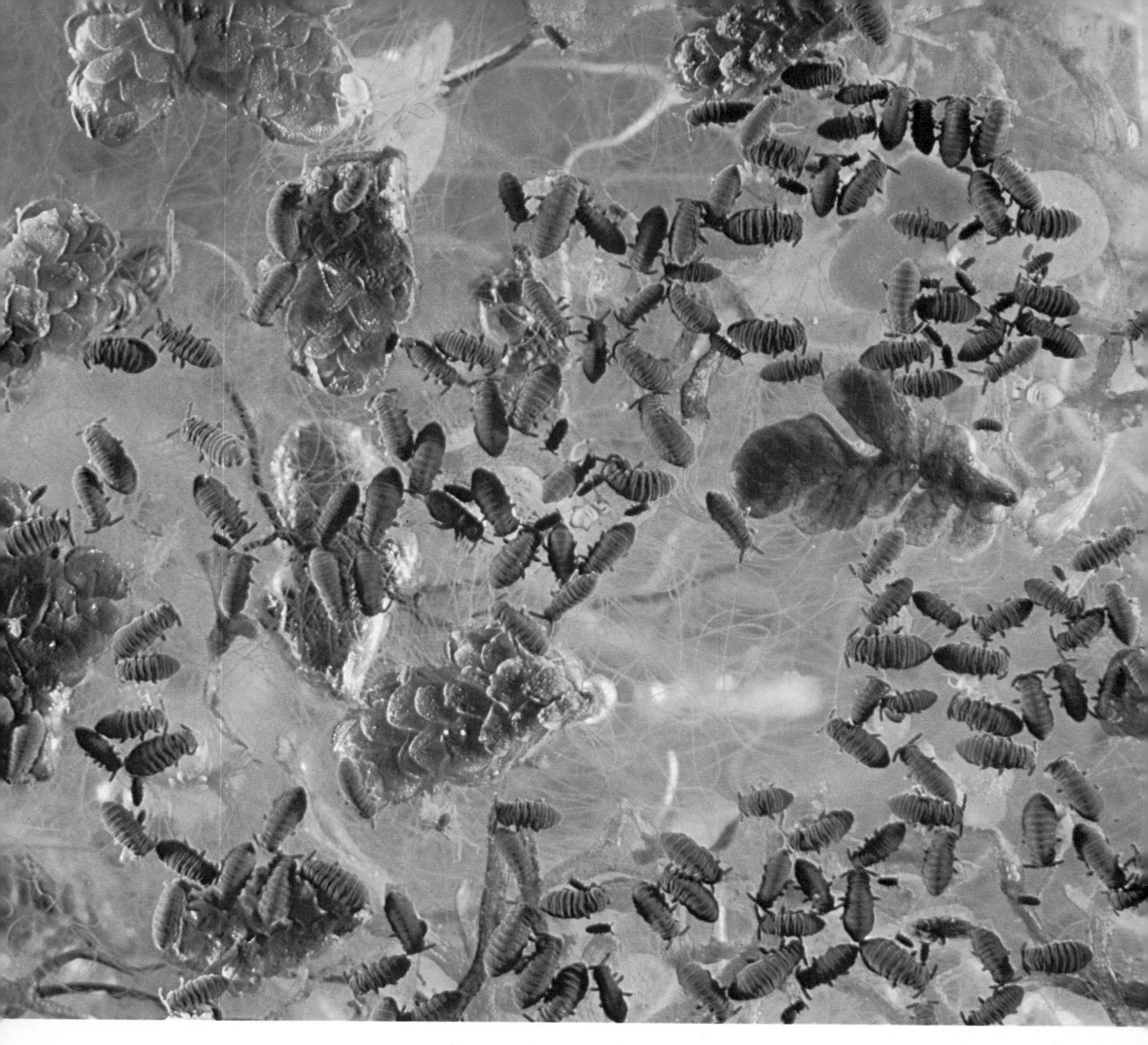

Above: water springtails cluster on the water surface to eat plant debris. These minute wingless insects can leap about with startling agility. This they do by flicking themselves upward by their tails. They can also crawl slowly over the water to forage.

invertebrates. Many can locate potential prey by recognizing the ripples or vibrations emanating from it as it struggles to escape from the surface tension. Like the other water bugs, to which they are closely related, the pond skaters have a proboscis for extracting the contents from the bodies of their prey. Different families of skaters can be found on the surfaces of still and running water, those in running water being generally smaller and stockier.

Another type of insect that is superbly adapted for life on the water surface is the black whirligig beetle. Its characteristic motion is more often recognized than the animal itself, for it lives up to its name by whirling constantly in crazy circles. Whereas some swimmers keep their left and right limbs in step, these creatures alternate them—hence the gyrations they perform. These beetles sometimes occur in large numbers, and they appear to gambol about in a high-speed game on a sunlit area of a ditch or pool. If you disturb the water, you simply increase the pace of their

activity; and some of them may even dive below the surface, where their shiny water-repellent covering protects them from being wetted and trapped. Their eyes have different sections for aerial and aquatic vision, and their middle and posterior pairs of legs are flattened for use as paddles. Whirligig beetle larvae are aquatic and live among submerged plants; when mature, they climb up these plants and build cocoons on the part growing above the water surface.

Whirligig beetles and water striders are typical winged insects, with stiffened forewings covering soft, membranous hindwings that are folded over the abdomen. Much less typical are the springtails—apparently rather primitive insects, whose ancestors, like themselves, never had wings. Some species of this widely distributed group live on and around the water. They are very small and look like mobile black dust sprinkled on the water. The name of the group is derived from their means of locomotion: the "tail" can be hooked under the body and the tension then quickly released, so that, as the tail hits the water, it flicks the whole animal into the air. Alternatively, a springtail can crawl slowly over the surface or down into the water, where it is protected by a covering of air trapped on the body. These tiny creatures feed on surface detritus.

The animals named above are continuously associated with fresh water, but we shall now have a look at a few whose links with the freshwater environment are less close, beginning with frogs and other amphibians. Amphibians were the first vertebrates to venture onto the land. They have never become totally adapted to a dry environment, but they are now found in a wide range of habitats, from the wholly aquatic to the completely terrestrial. The majority of amphibians can be divided into two groups: the newts and salamanders, and the frogs and toads. The first group still retain a tail for propulsion and, broadly speaking, they also spend more time in the water than most of the second group. Although most frogs and toads remain fairly close to the water, they have evolved an efficient means of locomotion on land through losing their tails and developing strong rear legs in the adult stage. Many salamander species are also terrestrial, living in such damp places as under logs.

Left: whirligig beetles gyrating on an Ontario pond. These small black beetles never collide, in spite of their frantic motions. They largely feed on insects that have fallen in the water.

Below: the edible frog lives in and near fresh water in lands from France to Russia. If its big eyes detect danger the frog dives and swims for cover with swift, powerful strokes of its long and immensely muscular hind legs.

Above: pleurodele newts courting. The sequence shows (left to right) a male below a female, with his front legs hooked over hers; the male swimming up with the female on his back; the male nudging the now receptive female.

Below: European common frogs mate in a pond among "clouds" of spawn. These frogs live largely on land, but like all amphibians they must start life in a moist environment. The male rides on the female's back to fertilize the eggs.

All amphibians have soft, relatively moist skins, through which they can get oxygen and obtain or lose water. Toads have tougher skin than the rest; they therefore lose less water and are better equipped for life on dry land. Male frogs and toads often croak to attract mates, but the amphibians with tails rely on coloration or scent to secure their females. The vivid colors of many frogs sometimes provide camouflage, and sometimes warn potential predators that the skin produces an unpleasant or even poisonous fluid. Certain South American Indian tribes tip their arrows with poison extracted from the skin of small frogs of a number of species, all brightly colored. These poisonous frogs do not need to seek security by changing the color of their skins to blend in with their habitats, as do their harmless relatives.

Frogs produce their eggs in a jellylike supporting substance. The mass of eggs has an easily recognizable shape that results from the movements of the male's hind legs as he rides on top of the female during the period when she is laying the eggs. Almost all amphibians lay their eggs in water, and on hatching, the first free-living stage is normally a small tadpole with external gills. These gills are lost when the internal respiratory system becomes functional. Limbs appear later, as the mature tadpole develops. Metamorphosis to the adult may involve many changes (including loss of the tail, as in the frogs), or few changes (perhaps nothing more than the development of lungs, in the tailed species). A few unusual amphibians retain their gills as adults; they are really more like the sexually mature larvae of other species than like most adult amphibians.

Almost all tadpoles are carnivorous, although some are plant-eaters in their earliest stage. Adult amphibians, like their young, are predatory; frogs, toads, and a few salamanders have sticky tongues with which to catch and hold their prey. Most amphibians that live mainly in the water find their prey among the detritus, whereas the more terrestrial forms eat a broad range of small invertebrates. The versatility of the amphibians has enabled them to colonize a wide variety of habitats all over the world, from the mountain brooks of Siberia to rivers in the caves of Yugoslavia to the South American jungles, where tiny frogs live their entire lives in the pools of water that collect in the "cups" formed by the base of the leaves of certain plants.

Although the "higher" vertebrates are not so

dependent on water as the amphibians, they must periodically visit rivers or lakes for water to drink. And so rivers and lakes are meeting places for large numbers of animals—a fact that is most obvious in localities such as the plains of Africa, where water is available from only a few sources. In the immediate vicinity of such a water supply the vegetation is more lush. The water holes of Africa are therefore a magnet to animals, because the animals can quench their thirst, and also because the establishment of a rich plant community provides food and shelter for herbivorous animals. The herbivores in turn offer a plentiful supply of food for predators, which feed on them as well as on the animals that are only temporary visitors to the water hole.

The predators that habitually frequent such areas include the crocodiles, alligators, and caymans. No member of these reptile groups is ever found very far from water. They rely on stealth, camouflage, and power to overcome their prey, which is usually less at home in the water than they are. All of them probably eat some fish as well as snakes, turtles, and frogs, but the larger species, at least, must rely mainly on birds and mammals for the bulk of their diet.

Crocodiles, alligators, and caimans, which all belong to the group we call the crocodilians, look

Right: raised eyes and nostrils enable a spectacled caiman to breathe and see while most of its body lies hidden under water. This helps it to ambush animals that come to drink. The raised eyes and nostrils of hippopotamuses and frogs are also adaptations to surface life.

Below: crocodiles, the world's heaviest reptiles, live mainly in water: they can even feed submerged. But they lay eggs on land, where the tough eggshells help to keep the contents moist. Thus crocodiles (along with other aquatic reptiles) are less closely tied to water than most amphibians.

very much alike. In fact, the best way to distinguish a crocodile from an alligator is to examine its teeth! When a crocodile's mouth is shut, the fourth tooth from the front in each half of the lower jaw projects outward and is visible; but its equivalent in an alligator fits into a socket of the upper jaw and cannot be seen. Alligators also have somewhat shorter and broader heads. Both groups contain large and small species. Largest of all is the estuarine crocodile of Southeast Asia, which may reach a length of 20 feet, the Nile crocodile of Africa (15 feet), the American crocodile (15 to 20 feet), and the American alligator (12 feet). No caiman reaches more than 15 feet, and that length is attained only by the black caiman, which (like all other caimans) is found in South and Middle America. It is not easy to distinguish between caimans and alligators on external characteristics. But internally they differ, for the caimans' nostrils are not divided and the belly plates are bony as well as those on the back.

All the crocodilians are well equipped for their amphibious life. Although ungainly they can move quite well on land, but they are at their best in water. They have webbed feet, which assist their mobility, but their major propulsive organ is the powerful tail. This also comes in handy for delivering defensive blows to enemies, and it is used in conjunction with the slashing teeth and jaws. In the water, the crocodilian may keep its eyes and nostrils—which are placed high on the head—above the surface, while the rest of the body is submerged. All crocodilians lay eggs on land near the water's edge or in nests, often made of rotting vegetation (in the process of decomposing, vegetation releases heat, which serves to incubate the eggs).

Hungry arrau turtles swarming in search of food. These freshwater reptiles grow to two feet in length and are themselves eaten as a delicacy by Amazon River Indians.

As has been pointed out, water provides a variety of "facilities," and a number of animals use these in several different ways. We have seen that water contains animals and plants of many sizes, and there are various ways to exploit this potential food supply. In the course of evolution over millions of years, a great variety of organisms have come to treat the water as their larder. And other creatures have found that the water can offer protection from their enemies.

To illustrate the first of these points, let us consider the gavial of India, a crocodilian that relies almost exclusively on water as a source of food. It has, so to speak, "given up" the chase after terrestrial prey. Instead, it feeds primarily on fish; and with its long, slender jaws, it is formidably equipped to catch them. It therefore remains in the water for almost its entire life, during

Above: a monitor lizard with a fish it has caught in Kenya's Lake Baringo. Several species of Old World lizards hunt in water.

The anaconda seldom strays far from water and becomes increasingly aquatic as it grows older. The biggest specimens spend almost all their time submerged.

which time the males may reach a length of 20 feet, although the average is 12 to 15 feet.

There are other reptiles that are mainly aquatic. Terrapins and some other turtles are amphibious, but many species stay in the water for most of the time, feeding on fish, small amphibians, and the larger aquatic invertebrate animals. They are partly herbivorous. Although most terrapins and turtles have hard shells, a few

Many birds have adapted to finding food in fresh water. Snakebirds (named for their long, sinuous necks) can dive, swim, and stab fish with their spearlike beaks.

A dipper gazes into a mountain stream (top), and plunges in after its prey (lower photograph). Waterproof feathers, nostril flaps, and extra eyelids fit dippers for diving.

have soft shells that consist of bony plates and skin. None of the freshwater turtles reaches the size of their marine relatives, but the alligator snapping turtle of the Mississippi River may grow to be six feet long, and is more savage than most species living in the sea. In defense, it can inflict crippling injury on a human being with its powerful hooked jaws.

Various snakes are found in and around fresh water in both the Old World and the New. The most common snakes belong to the colubrid group, and are mainly nonpoisonous. Some, known simply as water snakes, spend most of their time in the water, where they feed on fish and amphibians. But there are also some highly poisonous colubrid water-loving snakes, such as the North American water moccasin, and the water cobra that lives in Lake Tanganyika and

Waterfowl are forced to collect in shrinking areas of open water when ice covers northern lakes and rivers. Here, on an English reserve, Bewick's swans (white birds with black-tipped beaks) mingle with mute swans, Canada geese and other waterfowl.

various other nearby lakes of Central Africa.

The giant anacondas found throughout the tropical forests of South America are water-loving snakes, too, and the adults live almost completely in the water, feeding on suitably sized vertebrates that can be caught on land but dragged into the water before being eaten. Lastly, a unique snake can be found in Lake Bombon in the Philippine Islands—a sea snake that has become stranded in fresh water. This is the only known case of such a situation. As with all sea snakes, the bite of the Lake Bombon snake is extremely toxic.

We cannot leave the reptiles without mention of the basilisk, a species of lizard in the rain forests of Central and South America, that grows to a length of about 2½ feet including the tail. Though predatory and living near water, the basilisk does not appear to feed on aquatic organisms. It has, however, the singular ability of being able to run across the water surface when alarmed. It rises onto its back legs when moving at high speed—up to about five miles an hour has been recorded—and this enables it to dash over water for a few yards buoyed up by the surface film. In many parts of Latin America it is called the "Jesus Christ lizard."

But what of the animals that many people would mention first if asked to name some creatures that have an extremely close relationship with fresh water—birds such as ducks, geese, and swans? Many hundreds of bird families are largely if not entirely dependent on fresh water. They come to it not only to drink but also to feed on aquatic organisms or on the insects rising from the surface. And many build their nests in branches that hang above the water, where they are less accessible to their terrestrial enemies.

To begin with, let us consider a few oddities.

There is the dipper, which is the only true aquatic *passerine* (the zoological term for perching birds). Dippers can walk on a river bed or swim under water by using their wings: like penguins, they actually fly through the water. Many groups of seabirds also have occasional freshwater representatives. For instance, snakebirds (also known as darters) are related to cormorants; they sometimes spear fish on their sharp beaks while swimming below the surface of streams or lakes, then rise to the top, throw the fish upward, catch it as it falls, and swallow it in one movement. Another primarily marine group with a few species frequenting large areas of fresh water is the pelicans. Less agile than the snakebirds, they fish with their pendulous bills. Terns make up a further group of birds, most of which live near the seashore, but black terns and some other kinds can be found on rivers, lakes, and marshes in many parts of the world. One species, the Tibetan tern, spends its entire life on fresh water in the high valleys of the Himalaya.

The most familiar common water birds are undoubtedly the ducks. Although feeble on land, they fly well and are superb swimmers. The majority are dabbling ducks, which upend themselves and feed on plankton and debris just below the water surface. In contrast, the diving ducks swim down to the bed of a lake or river and feed on detritus or invertebrates. Fish are only a small part of the diet of most ducks, with the notable exception of the mergansers, which have toothed edges to their beaks, and are specialists at catching fish. Inside the beaks of other ducks are plates that filter organisms from a mouthful of water as it is expelled by the muscular tongue.

Among the most familiar dabblers are mallards, widgeon, and shovelers; and commonest among the divers are the pochards and tufted ducks. Two very familiar species, the mandarin and

Left: the great crested grebes of Eurasia, Africa, and Australasia have evolved a complex aquatic breeding ritual. Courtship rites include the penguin dance (beginning, near left) in which a pair of grebes face each other and rear up by treading water. Mating (below left) follows appropriately inviting gestures made by the male on his "cock's nest" of floating reeds (above left).

Right: a sacred kingfisher with a dragonfly. In spite of their name not all kingfishers catch fish, or even live near water. The large Old World group of forest kingfishers, to which the sacred kingfisher belongs, includes many birds that inhabit dry wooded country and prey upon insects and lizards. The fishing kingfishers form a separate group, with a worldwide distribution.

muscovy ducks, differ from the rest in having claws (in addition to webbing), and a strong hind toe, which enables them to perch in trees.

Swans are related to ducks. The great physical difference is the swan's longer neck. All northern swans are predominantly white and, except for the trumpeter, migratory; they breed mainly around the Arctic Circle but move farther south in winter. In the Southern Hemisphere dwell the large Australian black swan, and the South American black-necked swan. In general, although swans cannot dive, they swim well. Less aquatic than the swans are the geese, whose family can be divided into two branches, gray geese and black geese, which may belie their names by being neither truly gray nor truly black in color. All geese are remarkable for their ability to make long migrations flying in strict formation. The majority spend their time near coastlines, but some feed on plants and small invertebrates around fresh water.

Herons and their relatives differ in both shape and habits from the ducks, swans, and geese. The numerous species of heron are all alike in that they are large birds with relatively long necks, and many are primarily fish-eaters. They catch their prey by standing at, or slowly patrolling, the margins of a lake or river, then suddenly

An African jaçana and its baby seek insects and mollusks among the lily pads. Long toes spread each bird's weight across the floating leaves and prevent sinking. Jaçanas can spend their entire lives without touching land.

darting their heads into the water by extending their necks. They are mostly gregarious birds, which nest communally in heronries in tall trees. The nests themselves are often so close together that they may be separated by little more than the length of two extended necks. In contrast to the herons, the closely related bitterns are solitary birds. A male bittern establishes its territory prior to breeding, and the booming voice with which it advertises its presence is an unforgettable waterside sound. Bitterns are ground-nesting birds, but are difficult to see because of their brown or gray plumage—very unlike the blue and purple tints of the herons.

Ibises, too, belong to this group of wading birds characterized by long, featherless legs. The spectacular scarlet ibis of Central and South America is a tropical species, and feeds on many small freshwater invertebrates, especially crustaceans, as well as small fish. The flamingos enjoy a similar diet. They have unusual and very complex beaks, equipped to filter food from the water

A giant otter of Brazil munching an aquatic delicacy at the water's edge. Shallow water affords otters an easy hunting ground and enables them to drag their catch ashore.

Water offers many nesting birds protection from predators. This reed warbler rears its family among swamp plants where the helpless young are safe from land-based enemies.

that they pump in and out by moving their tongues backward and forward. With their long, slender legs and neck, and their delicate pink hues, flamingoes are certainly among the most graceful and beautiful of birds.

Among the other fish-eating birds that nest on or near fresh water are the loons—also called divers—which are weak fliers but excellent swimmers. What you are likely to notice if you see a loon on the water, usually in summertime, is its striking coloration. Much the same is true of the grebes, some of which have ornamental crests in addition to strong, contrasting body colors. The rails and gallinules are also brightly colored; they can easily be distinguished from the grebes, however, because they have much shorter, stubby beaks, suitable for feeding on their diet of plants and invertebrates.

Numerous terrestrial birds that cannot swim feed nonetheless on fish. The kingfishers dive into the water from overhanging branches. Ospreys and bald eagles hover above the water before plummeting down to secure their prey, which is sometimes very big. In all such cases, the birds may submerge to catch their food, but they remove it to the bank before eating it.

We cannot leave the water birds without mentioning the extraordinary jacana or lily-trotter.

Different species of this bird are found throughout the tropics, but they all have extremely long toes and claws, which enable them to run across floating leaves, collecting the seeds and small creatures that form their diet. Even their nests float, and may drift about in the wind and currents. A jacana may therefore spend its entire life on the water.

But what about the mammals? Many of them have become adapted to at least a semiaquatic way of life. Among the most totally dependent on feeding in water are the freshwater dolphins already described, the freshwater seal, and the otters, which have webbed hind feet and a rudder-like tail, and feed on fish. The freshwater seals that live around Lake Baikal in southern Siberia are in fact the only population of seals completely isolated from the sea. They look very similar to their marine relatives but they are much smaller. Lake Baikal is a deep lake that has been in existence for many years and has always suffered from a lack of nutrients. Now, however, an expanding timber industry is resulting in the addition of large amounts of material to the lake so that its waters are being enriched. Consequently, much of its fauna, including the seals, is threatened by the changing environment.

Many mammals use water not merely as a food source but as a means of protection for themselves and their young. Just as many birds, such as the reed warblers, construct their nests in plants that extend above the water, where they are less accessible to predators, mammals often have aquatic habits for the same reason. Thus, beavers construct their lodges on the dams they build across rivers or near the banks of lakes. Sometimes the lodges are very large and, with the entrance below water, virtually impregnable. Beavers do not feed in the water, though. They come ashore to eat shoots and the bark of trees—a diet that may be supplemented with a variety of fruits and vegetables.

The burrows of water voles, too, generally have their entrances concealed in the banks of rivers and streams, and are therefore well protected. In some situations, however, instead of building a burrow, a vole may build itself a nest from grass; such nests are safely suspended in reeds just above the water level. Water voles do feed partly in the water, on both plants and small animals, as well as on some terrestrial vegetation. Strangely, although they are so closely associated with water, they do not have webbed feet. Muskrats, which are much larger than voles, build their burrows in river or lake margins, and each burrow has two openings—one above and one below water level. The muskrats feed almost entirely on aquatic vegetation.

Two other interesting aquatic rodents are the coypu and the capybara, both native to South America. The coypu, whose fur is known as

Lake Baikal (far left), deepest and eighth largest of the world's lakes, lies in the remote center of the Asian landmass. Many unusual creatures have survived in its seclusion, among them the world's only freshwater seal (near left). Thousands of miles from its marine cousins, the Baikal seal nonetheless closely resembles them in most respects.

nutria, has been introduced, sometimes accidentally, into many temperate countries, where it occasionally causes trouble by destroying dikes or feeding on cultivated vegetables. It is strictly vegetarian and an excellent swimmer; a prolific breeder, it raises its young in burrows at the water's edge. The coypu, about 18 inches long, is by no means the largest living rodent. That honor belongs to the capybara, which may grow to four feet in length. This animal spends much of its time wallowing in pools, but it feeds on land, where it consumes a variety of plant material. Capybaras are related to, and look like, guinea pigs, except that they have much longer legs.

Perhaps the most peculiar of all freshwater mammals is the duck-billed platypus, an egg-laying mammal of Australia and Tasmania. It spends its life almost entirely in water, building a nest in the bank, with an underwater entrance that leads to an upward-sloping passageway. The main chamber is therefore above water level, and it is lined with grass, in which the female lays two eggs at a time. On hatching, the young suckle from their mother as other mammals do. The platypus uses its bill to collect detritus and small animals from the bed of the stream or lake in which it lives.

Let us finish this survey of secondarily aquatic life with a brief look at two other mammals, both African. These creatures are the hippopotamus and the sitatunga. The former is too well known to need much comment. Suffice it

Jaws agape, male hippopotamuses threaten each other with formidable teeth to win the possession of females. The hippo of Africa is the largest of the freshwater mammals.

to say that the reason why hippopotamuses spend most of their daylight hours dozing in water is that it not only cools their enormous bodies but also takes the weight off their legs. In the cool of the night they leave the water to forage for vegetation, sometimes at considerable distances from their daytime habitat. Whether on land or in the water, hippos are large and dangerous. Very different are the shy, retiring, and delicate sitatungas, which are semiaquatic antelopes that live in the swamps of tropical Africa. With their large, spreading feet, which keep them from sinking too far into the soft ooze, they are splendidly adapted to swamp life. Unlike the hippopotamuses, the main reason for their living in wet areas is probably security: they can outpace most predators in their unusual habitat.

It becomes evident from this consideration of animals that live on and around fresh water that our cycles of nutrients and other chemicals can be extended beyond the water onto land. Quite simply, we can say of many of the animals that live not *in* but *on* and *around* the water, that their removal of various elements of the aquatic flora and fauna partly balances the continuous input that we spoke of earlier. But there is also an added input from these animals themselves—from their feces and urine, and eventually from their dead bodies. They do not, therefore, disrupt the cycles; rather, they add a further dimension. If only the same thing could always be said about the species that remains to be discussed—man!

Its webbed feet, ducklike bill, and ability to lay eggs make the platypus the most peculiar of all freshwater mammals.

The rabbitlike lowland pacas of South and Central America are among the many rodents that live in river-bank burrows.

Man and Fresh Water

Man must drink fresh water to stay alive. This was true throughout the period of his evolution from some apelike ancestor, and it is the reason why it has been suggested that the earliest "true" men lived alongside lakes and rivers, probably in Africa. But it is not only in staying alive that man has found water useful. His whole history shows the influence of this liquid. When man survived by hunting, the area in which he lived must have centered on bodies of water to which

Modern man has transformed rivers such as the Thames, seen here at Battersea in London. Industrial buildings dominate once marshy, wooded banks; smoke darkens the sky; and man-made wastes taint the water. Such conditions sharply reduce the variety and quality of life that a river can hold.

his prey returned at regular intervals in order to drink. Later, when he learned to till the soil, he realized that water is vital to the growth of crops; the remains of old aqueducts and terraced hillsides for making maximum use of available water are evidence of his awareness of this need. And he learned that the aquatic environment itself can be exploited for food by fishing.

As increasingly large social groups made their appearance and began to establish permanent settlements, many of them settled close to fresh water. Rival communities fought for possession of desirable living space—and the most desirable areas were probably those that contained plenty of fresh water, and hence food. Trade between settled communities necessitates a means of transport, and water can serve this function in addition to quenching man's thirst and increasing his production of food. As we know from studies of the Middle East, and in particular of

the Nile Valley, in ancient times, boats or rafts linked human settlements for many centuries. The extent of these ancient civilizations' awareness of the immense value of water transport can be judged from the boats among the items entombed with the royalty of ancient Egypt.

For thousands of years, even up to the present day in some isolated communities, fresh water was used only for the above purposes and a few related ones, such as washing. Then came the Industrial Revolution and the beginning of great changes. Until that time, little more than 200 years ago, man's activities had very little disruptive effect on freshwater plant and animal life. His influence on the delicately balanced cycles and webs that comprise freshwater communities was no more destructive than that of any other animal. It is in the relatively short period since the middle of the 18th century that he has upset the natural tendency of the environment to remain in balance. This is what we mean when we speak of the problem of "pollution."

The development of industry started the ever-growing demand for energy, of which running water was an obvious source, and created a need for large quantities of water for industrial processes such as cleaning, cooling, and washing. An increasing population needing power for individual use and water to drink, the continued growth of industry to produce goods for that population—all this demands yet more water.

The energy of running water was initially harnessed through the waterwheel, which could be used to drive a wide variety of tools. Based on totally different principles but ultimately of greater importance was the steam engine—which depended, of course, on a liberal supply of fresh water. It may therefore be said that the Industrial Revolution was facilitated by water; and the importance of water in the generation of power has shown few signs of waning since the discovery of its use for driving electricity generators. Hydroelectric plants have played an important role in the 20th century, and will undoubtedly continue to do so in the 21st.

Running water has also been used to dispose of waste. It once seemed that an endless amount of waste, both organic and inorganic, could be carried to the sea without detrimental effects. Rivers have therefore been used as mobile refuse dumps, carrying progressively greater amounts of discarded material as they neared the coast.

If the uses of water lie in such activities, what problems have been found to result from them? First, damming rivers to harness water power may curb the flow of nutrients, thus reducing the life-supporting capability of the river's lower reaches. But the dumping process is more immediately harmful, and even the use of water in industrial processes or for cooling affects the

In the past, and in remote rural areas today, man's low numbers have let him use fresh water without disrupting freshwater life. Far Left: Egyptians in papyrus "reed" boats, in a mosaic from Roman times. Near left: Peruvian Indians still use reed boats.

environment. Cooling, for instance, depends on heat exchange—on the heating of water at the expense of the item being cooled. Thus, as the water returns to its source, less any lost as steam or by evaporation, it is hotter than when removed. The effect of the hot water is, in many respects, local, because it eventually cools to the temperature of the body of water from which it was separated. But near to the outlet that returns it to its source, the hot water must disturb the ecosystem, for some organisms can survive only between narrow temperature limits. The industrial use of water for cooling, which results in what is known as *thermal pollution*, may nevertheless be of negligible importance compared with other forms of pollution.

The most widespread form of water pollution is *chemical pollution*. This is a general term, and there are, in fact, different classes depending on such factors as the source, content, or effect. For example, sewage is a mixture of many substances, some of which are derived from human waste. Agricultural "runoff"—material that enters the water from farming land—is also a mixture: partly animal waste, but also chemicals such as surplus fertilizer, insecticide, weedkiller, and similar products that have been applied to the land for a specific purpose.

Sewage and animal wastes are fertilizers and, like the synthetic forms supplied nowadays by chemical industries, are equally efficient in fertilizing both land and water. As a rich source of plant nutrients, they are used to encourage the growth of crops on land; but in the water they primarily affect algae. We have already noted that algae have the potential to reproduce at great speed under suitable conditions, of which sunlight and an ample supply of nutrients are the most important. The addition of surplus nutrients therefore results in a massive increase in the algal population in summer, followed by an equally large decrease when the nutrients are exhausted or, with the coming of winter, the sun disappears. During both the increase and the decrease in algal population, the cycles in the water must contend with consequent extraordinary changes in some components. It is not surprising that a breakdown of the cycle may sometimes follow. For example, the massive algal population will feed and provide oxygen for a very large animal population, but when the algae die off, the animals may also die because of an inadequate supply of oxygen.

Moreover, although sewage and fertilizers overload parts of the cycle, other substances have totally opposite effects. Many insecticides and weedkillers are as effective in water as on land. Their entry into the water—particularly into lakes, where they may accumulate—can eliminate directly a number of essential steps in the nutrient cycles. Some industrial effluent, too, produces similar unfortunate effects. Industrial effluent is a term widely used to describe the waste, rubbish, and filth of many kinds resulting from industrial activity, which is released into water. And it need not be released directly; the air may act as an intermediary, with the substances dissolving into the water from a gaseous state or settling on the water's surface near their outlet.

The many kinds of industrial effluent affect the freshwater community in a number of ways. Substances that are poisonous can inhibit such life processes as the transmission of nerve impulses or the release of energy within an organism's body. Other types of effluent have indirect effects. In many industries, fine dusts are produced through the pulverizing of such minerals as limestone or chalk. When introduced into the water,

Above: man now heavily exploits rivers as moving highways for the cheap transport of raw materials and urban waste. The timber industry has long made use of rivers. This log jam covers a river near Terrace Bay on the northern shore of Lake Superior.

Above: floating rubbish, sewage, and industrial effluent transform this French river into a turbid open drain.

Below: these flyblown fish suffocated as the oxygen content of their water fell. Pollutants may cause such oxygen loss.

Above: products of modern industry often disfigure the countryside. Less obviously, they may release harmful chemicals into bodies of fresh water. In the case of these abandoned automobiles, gasoline and battery acid may leak into the river.

Right: French family picnicking by the Marne River in 1938. Since then, pollution has spoilt many waterside picnic sites. But people still prize the recreational value of clear mountain streams and tranquil, unpolluted waters.

Far right: concern about pollution has triggered campaigns to cleanse fouled and choked-up ponds, canals, and rivers, Here, conservationists rake surplus muck and rotting reeds from a village pond in south-east England.

this material may clog up various parts of an animal's body. For instance, it may get into the gills, where it coats their thin lining, reducing the rate at which oxygen can be absorbed and so leading to death by asphyxiation. This can occur in fish as well as in the many gilled invertebrates. Still other substances have unusual chemical effects in water. Some compounds attract other substances to themselves, and thus deplete the supply available as nutrients. Or the chemicals attracted may be not nutrients but substances that determine the acidity or the "hardness" of the water. Many organisms can live only within limited ranges of these factors and will die if exposed to unusual conditions. Thus, whichever way a specific effluent affects the environment, the effect on the aquatic community is either an overloading or a removal of some part of the complex cycles that sustain their life.

The polluted water that we are concerned about today, therefore, is not all polluted in the same way or to the same extent. "Pollution" is a blanket term that covers an immense variety of states produced by a range of polluting agents. And so there is no single answer to the problem; just as each kind of pollution may be different in origin, so each requires individual treatment. For instance, mass deaths of fish, periodically reported in the news media, may be due to the presence of a toxic chemical, whereas algal "blooms" and eutrophication are usually due to the rapid addition of nutrients. Thus, if we want to prevent recurrence of pollution in a body of water, we must find the source of the pollutant.

What about the fear of many environmentalists that fresh water is in increasingly short supply? Actually there is no global shortage of fresh water. But, although this bland statement may sound reassuring, it sidesteps the real problem; water is not always available where it is required, and in many places water shortage may indeed be becoming a serious problem. We have found substitutes for many commodities—for instance, plastics for metal and wood—but not for water. Water is in increasing demand for both industrial and domestic uses.

There *are* ways of augmenting the supply. We have already seen how fresh water is normally recycled by the forces of nature, and man is capable of speeding up the process by, for example, artificially causing rain to fall. But the possibilities are not endless, and activities such as these are very costly in both time and money. Fresh water should therefore be considered a precious substance available in only limited quantities through natural events.

As time passes, we are growing more and more aware of the problems of our position. Much has

been said and written about the two-pronged problem of water pollution and scarcity. Although the dangers may sometimes be exaggerated, the cures are certainly difficult. Once the cycles of life in fresh water are destroyed, it is not easy to reestablish them. The slow evolution of the freshwater community from marine and terrestrial ancestors and its adjustment to prevailing conditions has taken millions of years. It would therefore be unreasonable to expect the entire freshwater ecosystem to be able to adjust to immense new forces in a relatively short period. If we want to return our rivers and lakes as nearly as possible to the condition that they were in before the Industrial Revolution, it is essential to relieve them of more recent stresses and strains. It is true, as has been said, that the freshwater community can adapt to a changing environment. It can adjust to slow change by shifting the balance in different directions—which often results in the extinction of some groups of organisms and the acquisition of others. And so the extinction of species may not be due to man alone; in the future as in the past, it is to be expected that some animals and plants will become extinct through natural processes—in spite of man, as it were, rather than because of his activities. Yet there can be no doubt that man has done great damage, and continues to do so, to all the most basic of the natural resources.

The rivers and lakes of the world are too important to mankind to be allowed to become tragic victims of pollution. Tragedy for the freshwater community must inevitably mean tragedy for *us*. That is why it is appropriate that, having looked at many of the interesting and wonderful plants and animals that live in and around fresh water, we should finish up with a word about man. Water is an essential key to his survival, and the future of the lakes and rivers of the world may be the story of man himself. His ability—or his failure—to overcome the obvious problems confronting him today will inevitably be reflected in the state of the rivers and lakes in years to come. Water is not like a metal. We cannot exhaust our supply and then replace it with something else. Life is founded on water—it makes up four fifths of the human body. Nothing can substitute for it.

The world within the streams and lakes can serve as an example to us all, showing how we depend on our environment and upon each other, just as the fishes, insects, algae, and bacteria do. The living world of fresh water warrants all the study we can give it, because we can perhaps assure our own survival through an understanding of the principles on which the freshwater world is based.

A River System Modified by Man

Here is the river system shown on pages 36 and 37, this time after man has tapped its natural resources. The river course and the broad configuration of the landscape remain as they were. But much else has altered. The biggest single change has occurred on the flat land of the flood plain. There, seaborne goods shipped into the river mouth have nourished a sprawling factory city served by road, rail, air, canal, and river. Cables slung from pylons feed the city with electric current from a power plant near the large mountain lake. (This power plant harnesses the force of water pouring from the lake into a mountain valley out of sight beyond the lakeside cliff.) Mountain valleys carved by rivers and glaciers are too narrow to hold big urban centers. But skiing and boating facilities have helped to build the lakeside town as a vacation resort with a road link to the lowland city.

Index

Page numbers in *italics* refer to illustrations or captions to illustrations.

Picture Credits

Key to position of picture on page: (B) bottom, (C) center, (L) left, (R) right, (T) top; hence (BR) bottom right, (CL) center left etc.

Cover:	Hermann Eisenbeiss, München
Contents:	Photo Klaus Paysan, Stuttgart
9	Carl Perutz/Colorific!
10(L)	after Robert L. Usinger, *The Life of Rivers and Streams*, McGraw-Hill
10(R)	Ross/Jacana
11	Vala/Jacana
13	Photographed by Romano Cagnoni
14	after William H. Amos, *Life of the Pond*, McGraw-Hill
16(L)	Renaud/Jacana
17(L)	Jane Burton/Bruce Coleman Ltd.
17(R)	Jen and Des Bartlett/Bruce Coleman Inc.
18(L)	Heather Angel
19(L)	D. Oudet/Pitch
19(R)	P. Morris, Photographics
20	Ken Hoy/Ardea
21	Christian Cuny/Explorer
23	Adam Woolfitt/Susan Griggs Agency
25	Michael Freeman
26	Josef Muench
27	Munzig/Susan Griggs Agency
29	Colour Library International
31	John S. Flannery/Bruce Coleman Inc.
32(L)	Heather Angel
33	R. Everts/ZEFA
34	G. R. Roberts, Nelson, New Zealand
35(R)	Christine Foord/N.H.P.A.
39	L. West/Frank W. Lane
40(L)	G. R. Roberts, Nelson, New Zealand
41	Richard L. Fearn/Photo Researchers Inc.
43	Eric V. Grave/ Photo Researchers Inc.
45	Heather Angel
46(T)	James H. Robinson/Photo Researchers Inc.
46(B)	Jane Burton/Bruce Coleman Ltd.
49	D. Oudet/Pitch
50	Harald Mante/ZEFA
51(BL)	Hermann Eisenbeiss, München
51(R)	Heather Angel
53(T)	Oxford Scientific Films/Bruce Coleman Limited
53(B)	Heather Angel
55	after William A. Niering, *Life of the Marsh*, McGraw-Hill
56(L)	Michael Freeman
56(R)	John Markham/Bruce Coleman Ltd.
57(L)	Heather Angel
57(R)	Hermann Eisenbeiss, München
59	George Holton/Photo Researchers Inc.
61	Binois/Pitch
64(T)	Oxford Scientific Films/Bruce Coleman Ltd.
64(B)	Natural Science Photos
65(L)	Oxford Scientific Films/Bruce Coleman Ltd.
65(R)	Hermann Eisenbeiss, München
66	Heather Angel
68(L)	Jane Burton/Bruce Coleman Ltd.
68(R)	Oxford Scientific Films/Photo Trends
69(TL)	Philippe-Summ/Jacana
69(TR)	J. A. L. Cooke/Photo Trends
69(B)	Jane Burton/Bruce Coleman Ltd.
70	after William H. Amos, *Life of the Pond*, McGraw-Hill
72	Heather Angel
76(L)	Lynwood Chace/Frank W. Lane
77	P. H. Ward/Natural Science Photos
78(TL)	Heather Angel
78(BL)	W. Harstrick/ZEFA
79	John Goddard/Natural Science Photos
80	Jane Burton/Bruce Coleman Ltd.
81	Robin Fletcher/Natural Science Photos
83	Adam Woolfitt/Susan Griggs Agency
84(T)	P. Morris/Ardea
84(B)	Patricia Caulfield, Animals Animals © 1974
86, 87(T)	Jane Burton/Bruce Coleman Ltd.
87(BR)	Yvan Merlet/Frank W. Lane
88	Jane Burton/Bruce Coleman Ltd.
89(L)	Michael Freeman (Natural Science Photos
89(R)	Michael Freeman
92(TL)	Oxford Scientific Films/Photo Trends
92(BL)	Heather Angel
92(R)	Oxford Scientific Films/Bruce Coleman Limited
93	Heather Angel
94	Jane Burton/Bruce Coleman Ltd.
99	Ronald Thompson/Frank W. Lane
100(L)	M. Bellieud/Pitch
100(R)	J. A. L. Cooke/Photo Trends
101(TR)	Alan Blank/Bruce Coleman Inc.
101(BR)	Heather Angel
103(T)	Jane Burton/Bruce Coleman Ltd.
103(B)	Ken Brate/Photo Researchers Inc.
104(L)	Heather Angel
104	Tom Myers/FPG
106–7(B)	Oxford Scientific Films/Bruce Coleman Ltd.
107(T)	Heather Angel
111	Hermann Eisenbeiss, München
112	Heather Angel
113(B)	Norman R. Lightfoot/Photo Researchers Inc.
114–5(T)	Heather Angel
114(BL)	Hermann Eisenbeiss, München
115(B)	F. Greenaway/N.H.P.A.
116	Tom McHugh/Photo Researchers Inc.
117(T), 118	Michael Freeman
119(T)	Gerald Gubitt
119(B)	Bernard Stiegler/Jacana
120(L)	Trudy Unverhau, Animals Animals © 1974
120(R)	Oxford Scientific Films/Bruce Coleman Limited
121	Philippa Scott
122	M. Wiechmann/Frank W. Lane
123	Bruce Coleman Limited
124	G. D. Plage/Bruce Coleman Ltd.
125(L)	Michael Freeman
125(R)	John Markham/Bruce Coleman Ltd.
126	Novosti Press Agency
127	Sally Anne Thompson
128	Camera Press Ltd.
129(TR)	Douglas Baglin Photography Pty. Ltd./N.H.P.A.
129(BR)	Michael Freeman
131	Michael St. Maur Sheil/Susan Griggs Agency
132(L)	Mauro Pucciarelli, Rome
133	Fotofass
134(L)	G. R. Roberts, Nelson, New Zealand
135(TL)	Le Chapelain/Pitch
135(BL)	Fred Ward/Black Star, New York
135(R)	Alfred Eisenstaedt *Life* © Time Inc. 1974
136	Henri Cartier-Bresson/Magnum Photos
137(R)	Ford Motor Company

Artist Credits

© Aldus Books: Artist Partners (Roy Coombes) 36–7, 138–9, (Shireen Fairclough) 14–5; David Nockels 55, 108–9; Joyce Tuhill, Title page, 62–3, 70–1; Peter Warner 96–7.

We would like to thank Dr. Michael Hassell and Dr. Stuart McNeill of Imperial College, London, for providing the original sketch for the food web diagram on Pages 96–7.